Singular Optimal Control Problems

This is Volume 117 in
MATHEMATICS IN SCIENCE AND ENGINEERING
A Series of Monographs and Textbooks
Edited by RICHARD BELLMAN, *University of Southern California*

The complete listing of books in this series is available from the Publisher upon request.

Singular Optimal Control Problems

DAVID J. BELL
Department of Mathematics and Control Systems Centre
University of Manchester Institute of Science and Technology
Manchester, England

and

DAVID H. JACOBSON
National Research Institute for Mathematical Sciences
Council for Scientific and Industrial Research, South Africa
(Honorary Professor in the University of the Witwatersrand)

1975

Academic Press
London : New York : San Francisco
A Subsidiary of Harcourt Brace Jovanovich, Publishers

ACADEMIC PRESS INC. (LONDON) LTD.
24-28 Oval Road
London NW1

U.S. Edition published by
ACADEMIC PRESS INC.
111 Fifth Avenue
New York, New York 10003

Library of Congress Catalog Card Number: 75-21610
ISBN: 0-12-085060-5

Printed in Great Britain by Galliard (Printers) Ltd, Great Yarmouth

PREFACE

The modern theory of optimal control has its beginnings in the pioneering work of men such as Hohmann in Europe and Goddard in the United States. These men, and others like them, dreamt of the day when space travel would become a reality. It is the work of these pioneers which has helped to bring their dreams to partial fulfillment in such a brief period of time. The optimization problems of space flight dynamics such as minimum fuel and minimum time were quickly realized to be problems in the calculus of variations. It was thus that this very old and established branch of Mathematics received yet another new lease of life, a pattern which has been repeated since its birth in the seventeenth century.

Problems of high performance aircraft became important near the end of the Second World War. Maximum range of aircraft for a given quantity of fuel and minimum time to climb were typical optimization problems which arose. These too were clearly problems belonging to the same class as those from space dynamics although complicated by the presence of aerodynamic lift and drag.

Much of the mathematical theory for such problems had already been developed early in the twentieth century by Professor G. A. Bliss and his students at the University of Chicago. In particular, the optimization of aircraft and space vehicle flight paths for which the controls are finite and bounded was analysed by a technique described in a dissertation by F. A. Valentine, a research student of Bliss, in 1937. However, in 1959 L. S. Pontryagin presented his Maximum Principle which consolidated the theory for constrained problems.

Nevertheless, it soon became apparent that the mathematical theory available was not sufficient for certain special control problems in which the Pontryagin Principle yielded no additional information on the stationary control. These problems were described as singular problems and they have arisen in many engineering applications in fields other than Aerospace and also more recently in non-engineering areas such as Economics. In the last ten to twelve years much effort has been put into the development of new theory to deal with these singular problems. First, new necessary conditions were found for singular extremals to be candidate optimal arcs. Secondly, and more recently, new sufficient conditions and necessary and sufficient conditions have been found for such extremals to be optimal.

We feel that the time is now right for the theory of singular problems to be collected together, scattered as it is in numerous different journals, and presented under one cover. This is the purpose of the present volume. We are particularly pleased that this book should take its place alongside so many well-acclaimed texts in Richard Bellman's series which has proved its worth many times over. Our gratitude goes to all our colleagues and students, past and present, who have stimulated us both in our own researches. In particular, we would like to mention Y. C. Ho, D. Q. Mayne, J. L. Speyer, W. Vandervelde, D. F. Lawden, R. N. A. Plimmer and B. S. Goh. Finally, we acknowledge the help and encouragement received from Academic Press during the preparation of this book and our special thanks go to Mrs. G. M. McEwen and Mrs. M. E. Hughes who typed the manuscript.

D. J. BELL

June 1975 D. H. JACOBSON

CONTENTS

CHAPTER 1

An Historical Survey of Singular Control Problems

1.1 Introduction

It is well known that the fundamental problem of optimal control theory can be formulated as a problem of Bolza, Mayer or Lagrange. These three formulations are quite equivalent to one another (Bliss, 1946). We shall describe the Bolza problem since the accessory minimum problem, and the associated second variation with which we shall be much concerned in this book, appears as such a problem.

The problem of Bolza in optimal control theory is the following. Determine the control function $u(\cdot)$ which minimizes the cost functional

$$J = F[x(t_f),\ t_f] + \int_{t_o}^{t_f} L(x,u,t)\,dt \tag{1.1.1}$$

where the system equation is

$$\dot{x} = f(x,u,t) \tag{1.1.2}$$

subject to the constraints

$$x(t_o) = x_o \tag{1.1.3}$$

$$\psi[x(t_f),\ t_f] = 0 \tag{1.1.4}$$

$u(\cdot)$ is a member of the set U, t a member of $[t_o, t_f]$. (1.1.5)

Here x is an n-dimensional state vector and u is an m-dimensional control vector. The functions L and F are scalar and the terminal constraint function ψ is an s-dimensional column vector function of $x(t_f)$ at t_f. The functions L, F and ψ are assumed smooth. The set U is defined by

$$U \equiv \{u(\cdot) : u_i(\cdot) \text{ is piecewise continuous in time,}$$
$$|u_i(t)| < \infty, \qquad t_o \leq t \leq t_f,$$
$$i = 1, 2, \ldots, m\}. \tag{1.1.6}$$

The initial time t_o is given explicitly but the final time t_f may be unspecified.

The Hamiltonian for this problem is

$$H(x,u,\lambda,t) = L(x,u,t) + \lambda^T f(x,u,t) \tag{1.1.7}$$

and the following necessary conditions (Pontryagin's principle) hold along an optimal trajectory:

$$-\dot{\lambda} = H_x(\overline{x},\overline{u},\lambda,t) \tag{1.1.8}$$

$$\lambda(t_f) = F_x[\bar{x}(t_f), t_f] + \psi_x^T \nu \qquad (1.1.9)$$

$$H(t_f) = - F_t[\bar{x}(t_f), t_f] - \psi_t^T \nu \qquad (1.1.10)$$

where

$$\bar{u} = \arg \min_u H(\bar{x}, u, \lambda, t) \qquad (1.1.11)$$

$u(\cdot)$ a member of U.

Here $\bar{x}(\cdot)$, $\bar{u}(\cdot)$ denote the candidate state and control functions respectively, $\lambda(\cdot)$ denotes an n-dimensional vector of Lagrange multiplier functions of time, and ν is an s-dimensional vector of Lagrange multipliers associated with ψ.

A singular minimizing arc for the problem of Bolza is defined by Bliss (1946) as one for which the Legendre-Clebsch necessary condition is not satisfied with strict inequality. For the optimal control problem as formulated above the definition of Bliss is equivalent to the following. An extremal arc of the control problem is said to be singular if the $m \times m$ determinant $\det(H_{uu})$ vanishes at any point along it. Otherwise it is said to be nonsingular. In particular, if the Hamiltonian H is linear in one or more elements of the control function then the extremal is singular (Goh, 1966b).

The following definitions are used in the sequel:

Definition 1.1 Let u_k be an optimal singular element of the control vector u on the interval $[t_1, t_2]$ which appears linearly in the Hamiltonian. Let the 2q th time derivative of H_{u_k} be the lowest order total derivative in which u_k appears explicitly with a coefficient which is not identically zero on $[t_1, t_2]$. Then the integer q is called the order of the singular arc. The control variable u_k is referred to as a singular control.

Definition 1.2 Assuming all the elements $u_1, u_2, \ldots, u_m$ of the control vector u are singular simultaneously then u is called a totally singular control function when

$$H_u(\overline{x}, \lambda, t) = 0 \tag{1.1.12}$$

for all t in $[t_o, t_f]$.

Definition 1.3 A partially singular control function is one along which (1.1.12) holds for k subintervals of length T_i, i = 1, 2, ... , k and where

$\sum_{i=1}^{k} T_i < t_f - t_o$ (Jacobson, 1970b).

The concepts of total and partial singularity can be applied also to the accessory minimum problem in which the second variation of the cost functional (1.1.1) is to be minimized (see Sections 4.4 and 6.1).

In the important subclass of optimal control problems known as relaxed variational problems discussed by Steinberg (1971) some of the elements of the control vector u appear linearly and others appear nonlinearly.

In this book we shall be mainly concerned with problems which are totally singular for all t in $[t_o, t_f]$ (but see Section 5.4). In the partially singular case the nonsingular controls that are present can be eliminated via well-developed non-singular theory, leaving a problem totally singular in the remaining control variables (Robbins, 1967). The singular control theory to be discussed has been developed for and motivated by engineering problems. Nevertheless, singular problems may arise in any discipline where control theory is applied, evidence for which can be seen in economics (Dobell and Ho, 1967) and production from natural resources (Goh, 1969/70). In the following chapters necessary and sufficient conditions will be developed for singular optimal control of systems governed by ordinary differential equations. However, investigation is being carried out into singular control of discrete time systems (Tarn et al., 1971; Graham and D'Souza, 1970) and of systems with delay (Soliman and Ray, 1972; Connor, 1974).

1.2 Singular Control in Space Navigation

Practical problems involving singular controls arose early in the study of optimal trajectories for space manoeuvres. Trajectories for rocket propelled vehicles in which the thrust magnitude is bounded exhibit singularity in the rate of fuel consumption. Lawden (1963) formulates the fundamental problem of space navigation in the following way. $Ox_1x_2x_3$ is an inertial frame in which a space vehicle has position coordinates (x_1, x_2, x_3) and velocity components (v_1, v_2, v_3) at time t. The rocket thrust has direction cosines (ℓ_1, ℓ_2, ℓ_3) and the gravitational field has components (g_1, g_2, g_3). The mass rate of propellant consumption, bounded above by some finite constant $\bar{m}$, is denoted by m. If the rocket has an exhaust velocity c and the mass of the vehicle is M then its equations of motion are

$$\dot{v}_i = cm\ell_i/M + g_i(x_1, x_2, x_3, t) \tag{1.2.1}$$

$$\dot{x}_i = v_i, \qquad i = 1, 2, 3 \tag{1.2.2}$$

$$\dot{M} = - m. \tag{1.2.3}$$

The propellant consumption rate m must satisfy the inequality constraints

$$0 \le m \le \bar{m} \tag{1.2.4}$$

whilst the direction cosines may be written in the form

$$\ell_1 = \sin\theta \cos\phi, \quad \ell_2 = \sin\theta \sin\phi, \quad \ell_3 = \cos\theta \qquad (1.2.5)$$

where θ, ϕ are spherical polar coordinates. It is required to choose the control variables $m(\cdot)$, $\theta(\cdot)$, $\phi(\cdot)$ in order to minimize the fuel used in transferring the vehicle between two given positions at each of which the vehicle's velocity is specified. The final time t_f may or may not be given explicitly. State variables $x_i(\cdot)$, $v_i(\cdot)$, $M(\cdot)$ must satisfy boundary conditions

$$x_i(t_o) = x_{io}, \quad v_i(t_o) = v_{io}, \quad M(t_o) = M_o \qquad (1.2.6)$$

$$x_i(t_f) = x_{if}, \quad v_i(t_f) = v_{if} \qquad (1.2.7)$$

and the cost functional can be written as

$$J = -M(t_f). \qquad (1.2.8)$$

In this problem the Hamiltonian is linear in the rate of fuel consumption and this control variable turns out to be a singular control.

During the 1950's it was not known whether a given manoeuvre in space using a small, continuous thrust would be more economical in fuel than the previously accepted optimal procedure of impulsive boosts (Hohmann transfers). Some numerical results (Forbes, 1950) suggested that in certain circumstances less fuel would be used in an orbital transfer by following a spiral path using a small but continuous thrust throughout the maneouvre than by a Hohmann transfer. Other analysis (Lawden, 1950, 1952) appeared to show that the so-called intermediate-thrust arcs (along which the fuel expenditure rate is non-zero but less than $\overline{m}$) were inadmissible in a fuel optimal trajectory. The true status of the intermediate-thrust arcs remained hidden until the mathematical theory had been further developed (see Section 1.5.1 and Chapter 3). Much of the early work in aerospace has been surveyed elsewhere (Bell, 1968).

1.3 Method of Miele via Green's Theorem

A method which deals successfully with problems involving singular controls is due to Miele (1950-51). It is based upon a transformation using Green's theorem relating line and surface integrals. As developed by Miele the method is applicable only to a particular class of linear problems in two dimensions. The problems are linear in the sense that the cost functional and any isoperimetric constraint are linear in the derivative of the dependent variable. The general cost functional in this class of problems can be written as

$$J = \int_{x_o}^{x_f} [\phi(x,y) + \psi(x,y)\,dy/dx]dx \tag{1.3.1}$$

where ϕ and ψ are known functions of x and y. An admissible region in (x,y)-space is determined from the constraints in the problem formulation and the optimal trajectories within this region are determined by the sign of the function

$$\omega(x,y) = \frac{\partial\psi}{\partial x} - \frac{\partial\phi}{\partial y} . \tag{1.3.2}$$

If this function changes sign within the admissible region then a typical trajectory will contain two bang-bang arcs separated by a singular arc, the latter being defined by the equation

$$\omega(x,y) = 0. \tag{1.3.3}$$

Many problems in flight mechanics can be formulated in linear form of the above type provided suitable assumptions are made. Such problems are discussed by Miele (1955a, 1955b). The general theory of the extremization of linear integrals by Green's theorem for the two-dimensional case has been presented together with a number of applications to aerospace systems by Miele (1962). The synthesis of the optimal control by Miele's method including a demonstration of uniqueness of the singular control and the determination of the optimal strategy has been discussed by Hermes and Haynes (1963). The generalized Stokes's theorem of the exterior calculus has enabled Haynes (1966) to extend Miele's method to n dimensions.

1.4 Linear Systems - Quadratic Cost

Singular control for time-invariant, linear systems with quadratic cost function, free terminal time and fixed endpoint has been studied quite extensively. In particular, Wonham and Johnson (1964) and Bass and Webber (1965) have studied the system

$$\dot{x} = Ax + Bu \tag{1.4.1}$$

$$|u_i| \leq 1, \qquad i = 1, 2, \ldots, m \tag{1.4.2}$$

with cost function

$$J = \int_{t_o}^{t_f} x^T Q x \, dt \tag{1.4.3}$$

under the assumptions i) u is a scalar, ii) matrices A, B and Q are constant, iii) Q is positive definite. They have shown that the optimal control $\bar{u}$ is of the form

$$\bar{u} = k^T x \tag{1.4.4}$$

on the singular strip

$$|k^T x| \leq 1 \ , \quad k_1^T x \equiv 0 \tag{1.4.5}$$

where k and k_1 are constant vectors. The singular control is operative within the region of state space defined by this strip. Johnson and Gibson (1963), Athans and Canon (1964) have carried out associated work. In the latter reference a cost function is taken in the form

$$J = \int_{t_o}^{t_f} (k + |u(t)|)dt \tag{1.4.6}$$

with k a positive constant and u(t) a scalar satisfying $|u| \leq 1$. It is found that optimal singular controls exist if $k \leq 1$ but not otherwise. In all the references quoted above in this section the methods of analysis are rather special and are not directly applicable to time-varying systems with non-positive definite performance weighting matrices.

An early paper discussing the time-varying problem was by Johnson (1965). Assumptions i) and ii) mentioned above were relaxed in a paper by Rohrer and Sobral (1966) whereas results obtained by Sirisena (1968) follow from a relaxation of iii). An analysis in which all three assumptions were relaxed was given by Moore (1969) and Moylan and Moore (1971). An excellent book which discusses the topics of this section is the one by Anderson and Moore (1971).

1.5 Necessary Conditions for Singular Optimal Control

1.5.1 The Generalized Legendre-Clebsch Condition

In 1959 Leitmann suggested in private correspondence that the intermediate thrust (IT) arcs arising from minimum fuel problems of space navigation may after all be candidate sub-arcs for optimal trajectories. The first development following this suggestion was the discovery of the form of these IT-arcs in two dimensions and in an inverse square law gravitational field (Lawden, 1961, 1962, 1963). It was found that if the time of transit is not predetermined then the IT-arc is a spiral with its pole at the centre of attraction, a trajectory which has since become known as Lawden's spiral. The status of general IT-arcs in a central, time-invariant force field is given by Archenti and Vinh (1973).

The difficulty at this stage of the investigations was that the classical Legendre-Clebsch condition

$$\frac{\partial}{\partial u} H_u \geq 0 \tag{1.5.1}$$

which yields further information for nonsingular problems is trivially satisfied along a totally singular arc. More and more researchers thus became interested in these problems of singular control (e.g. Snow, 1964; Hermes, 1964) and particularly in the search for further necessary conditions which would help to establish the status of singular trajectories such as Lawden's spiral. In order to deduce new conditions for optimality it was necessary to study the second variation of the cost functional which must be non-negative for a minimum value of that functional. In certain cases it is possible to use the second variation directly in order to establish the optimality or otherwise of both nonsingular (Bell, 1965) and singular arcs (Bell, 1971a). However, the main effort during the last decade has been to deduce new necessary, sufficient, and necessary and sufficient conditions for singular problems from the second variation. A full discussion of the second variation will be found in Chapters 2 and 4.

When the control is a scalar, Kelley (1964a) deduced a new necessary condition for singular optimal control by studying the second variation under a special control variation. Kelley's method was generalized by Tait (1965), Kopp and Moyer (1965), Kelley et al. (1967) to give what has since become known as the generalized Legendre-Clebsch condition (Kelley-Contensou test):-

$$(-1)^q \frac{\partial}{\partial u} \frac{d^{2q}}{dt^{2q}} H_u(\bar{x},\lambda,t) \geq 0 \qquad (1.5.2)$$

where the integer q is the order of the singular problem.

The generalized Legendre-Clebsch condition for a vector control was obtained by Robbins (1967) and Goh (1966b). In this case the controls can appear in an odd time derivative of H_u but if this does occur then there necessarily exists a control u in U such that the second variation is negative (for a minimization problem). Hence, the generalized Legendre-Clebsch condition for vector control is

$$\frac{\partial}{\partial u}\frac{d^p}{dt^p}H_u = 0 \quad \text{for all } t \text{ in } [t_o, t_f] \quad p \text{ odd} \tag{1.5.3}$$

and

$$(-1)^q \frac{\partial}{\partial u}\frac{d^{2q}}{dt^{2q}}H_u \geq 0 \quad \text{for all } t \text{ in } [t_o, t_f] \tag{1.5.4}$$

assuming that the matrices H_{xx}, H_{ux}, f_x and f_u are sufficiently differentiable with respect to time. For p = 1, eqn(1.5.3) implies $H_{ux}f_u$ is symmetric, a powerful necessary condition in the case of vector controls (Speyer, 1973). On the other hand, eqn (1.5.3) is trivially satisfied when u is a scalar. Incidentally, the result of Speyer (1973) is contrary to Schultz and Zagalsky (1972) but consistent with the findings of Bryson et al. (1969).

The generalized Legendre-Clebsch condition or its equivalent was used by Robbins (1965) and by Kopp and

Moyer (1965) to test the optimality of Lawden's singular spiral, a problem of order q = 2. The new condition was not satisfied and so the spiral cannot form part of a minimum fuel trajectory. The same result was obtained by other workers using a special variation in the control (Keller, 1964) and by canonical transformations of the Hamiltonian (Fraeijs de Veubeke, 1965).

An alternative approach to singular problems was by a transformation on the state and control variables (Kelley 1964b; Speyer and Jacobson, 1971). Under this transformation a state space of reduced dimension is obtained and the singular problem (hopefully) changed to a nonsingular one in which the classical Legendre-Clebsch condition can be applied. If this is not the case then a further transformation to a state space of even smaller dimension is indicated. This method was used to investigate Lawden's spiral (Kelley, 1963). A generalization of Kelley's transformation has been given by Mayne (1970).

The general theory for both the generalized Legendre-Clebsch condition and Kelley's transformation has been given by Kelley et al. (1967) including the application of these theories to Lawden's spiral. In this last reference (amongst others) a plausibility argument was put forward concerning the satisfaction of terminal conditions associated with the second variation. It was suggested that the special control variations used in the derivation of the generalized Legendre-Clebsch condition could be altered by the addition of appropriate functions of time. Such additional functions would be of an order of magnitude which allowed the terminal conditions to be satisfied exactly but would not affect the sign of the second variation. However, it is sometimes possible to generate special state variations which satisfy both the equations of variation and the terminal conditions. This has been done in the case of Lawden's spiral (Bell, 1971b).

The existence of sufficient control variations to enable one to cancel out undesirable low order effects of the special control variations, as discussed above, is equivalent to a normality assumption. No such assumption is necessary in the Pontryagin Principle and recently this Principle has been extended to a so-called High Order Maximal Principle (Krener, 1973). This new Principle includes the generalized Legendre-Clebsch and Jacobson (see Section 1.5.2 below) conditions when they apply, together with other conditions, without the assumption of normality.

The difficulty caused by the fact that singular extremals may be transformed into singular extremals under a single application of Kelley's transformation was investigated by Goh (1966a) using a procedure by which the generalized Legendre-Clebsch condition for singular extremals can be deduced and, unlike Kelley's transformation, retains the full dimensionality of the original problem. It was again proved that Lawden's spiral is non-optimal and furthermore IT-arcs are candidates for an optimal trajectory only if the thrust possesses a component which is inwardly directed to the centre of attraction, a result which had already been noted by Robbins (1965). The transformed accessory minimum problem discussed by Goh was also used to further develop the generalized Legendre-Clebsch necessary condition for vector control problems (Goh, 1966b). This new set of conditions was applied to several problems including a doubly singular problem in interplanetary guidance (Breakwell, 1965) and to a class of identically singular problems, certain members of which had already been studied by Haynes (1966) using an extension of Miele's method discussed above in Section 1.3. The methods of Kelley and Goh have also been applied to a singular arc arising from the problem of a stirred tank reactor in the field of Process Control (Bell, 1969).

A number of investigations into singular problems have also been carried out by Russian workers (Bolonkin, 1969; Gabasov, 1968, 1969; Gurman, 1967; Vapnyarskiy, 1966, 1967). In particular, Gabasov (1968, 1969) obtained a slightly stronger version of a second necessary condition which is discussed in the following section. A comparison of the results of

Vapnyarskiy (1967) and Bolonkin (1969) with those of Goh (1966b) is given by Goh (1973).

Once the generalized Legendre-Clebsch condition had been established it was possible to lay down a procedure for the derivation of the singular extremals. This was done for the linear, time-varying system with quadratic cost functional by Goh (1967).

1.5.2 The Jacobson Condition

Having established the generalized Legendre-Clebsch condition researchers began to look for generalizations of the known sufficient conditions for the nonsingular problem. However, before such a generalization was found, a new necessary condition for non-negativity of the singular second variation, not equivalent to the generalized Legendre-Clebsch condition, was derived (Jacobson, 1969, 1970b). This new condition, known as Jacobson's condition, is as follows.

$$H_{ux}f_u + f_u^T Q f_u \geq 0 \qquad (1.5.5)$$

where

$$-\dot{Q} = H_{xx} + f_x^T Q + Q f_x \qquad (1.5.6)$$

$$Q(t_f) = F_{xx}[\bar{x}(t_f), t_f] \qquad (1.5.7)$$

and the partial derivatives f_u, H_{ux} and f_x are all evaluated along the singular arc $\bar{x}(\cdot)$, $\bar{u}(\cdot)$. A strong version of this condition is given by Mayne (1973).

In the classical calculus of variations problem the Jacobson condition can be replaced by an equivalent necessary condition on the terminal point of the singular extremal (Goh, B.S. On Jacobson's Necessary Condition for Singular Extremals. Unpublished Research Note). A generalization of this end condition leads to a result originally given by Mancill (1950). It has been shown by Jacobson (1970a) that in general the generalized Legendre-Clebsch condition and the Jacobson condition together are not sufficient for optimality.

1.6 Sufficient Conditions and Necessary and Sufficient Conditions for Optimality

A sufficient condition for a weak local minimum in a nonsingular problem is that the second variation be strongly positive (Gelfand and Fomin, 1963). This condition gives rise to a well known matrix Riccati differential equation. In singular problems the second variation cannot be strongly positive (Tait, 1965; Johansen, 1966) but investigations have shown that Riccati-type conditions do exist for the singular case. Jacobson (1970b), using a direct approach, derived sufficient conditions for the second variation to be non-negative in both singular and nonsingular control problems. These conditions are that there exist a real symmetric bounded, matrix function of time $P(\cdot)$ such that

$$H_{ux} + f_u^T P = 0 \quad \text{for all } t \text{ in } [t_o, t_f] \tag{1.6.1}$$

$$\dot{P} + H_{xx} + f_x^T P + P f_x \geq 0$$

$$\text{for all } t \text{ in } [t_o, t_f] \tag{1.6.2}$$

and

$$Z^T [F_{xx} + \nu^T \psi_{xx} - P]_{t_f} Z \geq 0 \tag{1.6.3}$$

where Z is the $n \times (n-s)$ matrix

$$Z = \begin{pmatrix} -D_1^{-1} D_2 \\ \hline I \end{pmatrix} \tag{1.6.4}$$

and the $s \times s$ matrix D_1 and the $s \times (n-s)$ matrix D_2 are such that

$$\psi_x = (D_1 \; D_2). \tag{1.6.5}$$

The above conditions in strengthened form are sufficient conditions for a weak relative minimum. It is demonstrated that the two necessary conditions for singular optimal control, discussed in Section 1.5 above, are implied by the new conditions for the case of totally singular control and unconstrained terminal state. Related to these sufficient conditions of Jacobson, with the matrix ψ_x the null matrix, are conditions found in the network theory literature in connection with passivity of electrical networks; see Rohrer and Sobral (1966), Rohrer (1968), Silverman (1968), Anderson and Moore (1968a, 1968b), Moore and Anderson (1968). The sufficient conditions (1.6.1-3) are very similar in form to certain necessary and sufficient conditions for positive real matrices (Anderson, 1967) which suggested that perhaps Jacobson's conditions may also be necessary. An alternative approach to the derivation of sufficient conditions for non-negativity of the second variation, in which the generalized Legendre-Clebsch condition (1.5.2) with $q = 1$ is satisfied with strict inequality and $\psi \equiv 0$, is by the transformation method of Goh or Kelley.

The approach used by Goh transforms the singular accessory minimum problem (AMP) into a nonsingular, non-classical one. The transformed problem is non-classical in that the control variations appear in the boundary conditions and outside of the integral in the second variation. Nevertheless, for certain classes of totally singular problems the transformed AMP has been used by McDanell and Powers (1970) to develop a sufficient condition which includes a Jacobi-type condition in the form of a matrix Riccati equation. A weakened form of this sufficiency condition is shown to be necessary for a smaller class of singular problems. The results of McDanell and Powers have been generalized by Goh (1970) and Speyer and Jacobson (1971).

The transformation proposed by Kelley (1964b) has been applied to the AMP of a totally singular optimal control problem (Speyer and Jacobson, 1971). The transformation can be obtained in closed form and the resulting problem is nonsingular. The generalized Legendre-Clebsch condition and Jacobson's necessary condition (for the unconstrained problem) are identified without the need for special variations. The results obtained are similar to those of Goh, McDanell and Powers mentioned above. Eqns(1.6.1-3) in strong form (strict inequality) and with unconstrained terminal state ($\psi \equiv 0$, $Z = I$), applied to the transformed AMP, are equivalent to the generalized Legendre-Clebsch condition and to the Jacobi condition (matrix Riccati equation) for the transformed nonsingular problem (Jacobson, 1970c). The sufficiency conditions of Jacobson are thus seen to be both necessary and sufficient. Since the transformation is nonsingular and Jacobson's conditions are coordinate independent these conditions are necessary and sufficient for optimality of singular problems in the original state space. This result is true for first-order arcs and for problems with or without terminal state constraints. If the problem is singular of order higher than one then the reduction of the state space to achieve a nonsingular problem requires repeated application of the transformation technique. This is cumbersome especially if there are multiple control variables. Furthermore, the method requires the coefficients of the time dependent variation to be many times differentiable.

Jacobson's sufficiency conditions can also be shown to be necessary by a limit approach (Jacobson and Speyer, 1971) which avoids the above difficulties. Here the singular second variation is made nonsingular by the addition of a quadratic functional of the controls:

$$\frac{1}{2\varepsilon}\int \delta u^T \delta u \; dt.$$

By allowing $\varepsilon \to \infty$ the optimality conditions for the singular problem are deduced from the limiting optimality conditions of the synthesized nonsingular second variation. This limit approach had been used previously (Jacobson et al., 1970) as a computation technique. The sufficiency conditions of Jacobson (1970b) are again shown to be necessary for non-negativity of the singular variation. Indeed, these conditions are shown to be necessary and sufficient for a weak minimum for a class of singular control problems and in certain cases a strong minimum is implied. In this limit approach a direct proof of necessary and sufficient conditions for optimality is obtained without the need to transform the problem to a reduced state space. Moreover, the concept of order of the singular arc is not required in the proof and differentiability requirements are not as severe as those demanded by Speyer and Jacobson (1971). In particular, the strong generalized Legendre-Clebsch condition is relaxed and a slightly more abstract version of condition (1.6.3) and condition (1.6.6) below is obtained. This set of conditions is considered in Section 4.6, Chapter 4 (Theorem 4.5). They have been extended to the partially singular case by Anderson (1973).

Following on immediately from the results discussed in the last two paragraphs, a general sufficiency theorem for non-negativity of a large class of second variations was presented by Jacobson (1971a) for the partially singular case. It is that there should exist for all t in $[t_o, t_f]$ a continuously differentiable, symmetric, matrix function of time $P(\cdot)$ such that

$$\begin{bmatrix} \dot{P} + H_{xx} + Pf_x + f_x^T P & H_{xu} + Pf_u \\ H_{ux} + f_u^T P & H_{uu} \end{bmatrix} \geq 0 \quad (1.6.6)$$

for all t in $[t_o, t_f]$, together with (1.6.3-5). A proof of this condition in the case of unconstrained terminal states is given in Section 4.5 of Chapter 4 (Theorem 4.2). Jacobson then applied this condition to both the totally singular case and to the nonsingular case. Sufficient conditions were thus obtained for the two special cases, demonstrating that both singular and nonsingular second variations can be treated in a common general framework. The results developed by Jacobson and Speyer through the transformation approach and the limit approach lead to the conclusion that the sufficiency conditions are also necessary for optimality for a large class of problems. In the nonsingular case the well known Riccati differential equation emerges and, since this is known to be a necessary condition for the nonsingular second variation, this implies that the sufficiency conditions of the theorem are also necessary. In the singular case the algebraic and differential inequalities (1.6.1-3) in strengthened form are obtained and these have been proved necessary and sufficient conditions for optimality in the 1971 papers of Jacobson and Speyer. Thus, in the singular case, the sufficiency conditions are also necessary. Assuming that the generalized Legendre-Clebsch condition is satisfied with strict inequality, a differential equation of Riccati type is obtained which implies and is implied by (1.6.1-3); see Section 4.5, Chapter 4

(Theorem 4.4).

Further results arising from the investigations of Jacobson and Speyer are obtained for a class of quadratic minimization problems whose optimal control functions are partially singular (Jacobson, 1972). An explicit expression for the time-varying, singular surface (hyperplane) is obtained. Sufficient conditions for totally bang-bang arcs and for totally singular arcs are matched for the case of a first-order singular problem to yield sufficient conditions for optimality of partially singular problems. The investigations do not include terminal constraints but do allow non-positive definite, time-varying, weighting matrices in the performance criterion and fixed terminal time. They therefore contribute to the understanding of problems discussed in Section 1.4 above.

An important paper by Moyer (1973) derives the conditions guaranteeing that a singular extremal which joins fixed end points provides a strong minimum for the independent time variable. Under proper assumptions this paper extends the Weierstrass concept of a field. Another interesting method for singular problems has been suggested by Vinter (1974). Instead of modifying the cost functional as is done by Jacobson and Speyer (1971) an alternative approach is to modify the control constraint set. The method is applied to a class of linear systems and again the solution of the modified problem approximates the solution of the original problem to any required accuracy. Sufficient conditions for the singular problem have also been derived by Mayne (1973) using differential dynamic programming.

The mathematical characterization of optimal controls which contain both singular and nonsingular arcs is still not complete. Preliminary results on the matching of singular and nonsingular arcs at junction points were obtained by Kelley et al., (1967) and by Johnson (1965). Further work clarified and extended these results (McDanell and Powers, 1971). The main result of this latter paper is that the sum of order of the singular arc and lowest order time derivative of control which is discontinuous at the junction must be an odd integer when the strengthened Legendre-Clebsch condition is satisfied. New necessary conditions which do not require an analyticity assumption are developed. These aid in characterizing problems which may possess non-analytic junctions. Also, McDanell and Powers (1971) have shown that a continuous junction for a singular arc of odd order is realizable contrary to a result given by Kelley et al., (1967). Results from two previous papers (McDanell and Powers, 1970, 1971) are used to investigate the local switching behaviour of the singular guidance problem associated with the Saturn VAS-502 flight (Powers and McDanell, 1971). A numerical investigation is carried out to obtain a satisfactory suboptimal guidance scheme using the computational procedure suggested by Jacobson et al. (1970).

Outlines of the methods of proof of the necessary and sufficient conditions mentioned above are given in the survey paper by Jacobson (1971b). Other survey papers on the singular control problem have been written by Gabasov et al. (1971) and Gabasov and Kirillova (1972), in which more references to the Russian literature will be found, and by Marchal (1973) which discusses the two different meanings of the term 'chattering' in optimization theory.

References

Anderson, B. D. O. (1967). A System Theory Criterion for Positive Real Matrices, SIAM J. Control 5, 171-182.

Anderson, B. D. O. (1973). Partially Singular Linear-Quadratic Control Problems, IEEE Trans. autom. Control AC-18, 407-409.

Anderson, B. D. O. and Moore, J. B. (1968a). Network Realizations of Time-Varying Passive Impedances, Tech. Rep. EE-6810, Univ. Newcastle, Australia.

Anderson, B. D. O. and Moore, J. B. (1968b). Extensions of Quadratic Minimization Theory. II Infinite Time Results, Int. J. Control 7, 473-480.

Anderson, B. D. O. and Moore, J. B. (1971). "Linear Optimal Control", Prentice-Hall, Englewood Cliffs, N.J.

Archenti, A. R. and Vinh, N. X. (1973). Intermediate-Thrust Arcs and their Optimality in a Central, Time-Invariant Force Field, J. Opt. Th. Applic. 11, 293-304.

Athans, M. and Canon, M. D. (1964). On the Fuel-Optimal Singular Control of Nonlinear Second-Order Systems, IEEE Trans. autom. Control AC-9, 360-370.

Bass, R. W. and Webber, R. F. (1965). On Synthesis of Optimal Bang-Bang Feedback Control Systems with Quadratic Performance Index, Proc. 6th JACC, 213-219.

Bell, D. J. (1965). Optimal Trajectories and the Accessory Minimum Problem, Aeronaut. Q. 16, 205-220.

Bell, D. J. (1968). Optimal Space Trajectories - A Review of Published Work, Aero. Jl. R. aeronaut. Soc. 72, 141-146.

Bell, D. J. (1969). Singular Extremals in the Control of a Stirred Reactor, Chem. Engng. Sci. 24, 521-525.

Bell, D. J. (1971a). The Second Variation and Singular Space Trajectories, Int. J. Control 14, 697-703.

Bell, D. J. (1971b). The Non-Optimality of Lawden's Spiral, Astronautica Acta 16, 317-324.

Bliss, G. A. (1946). "Lectures on the Calculus of Variations". Univ. Chicago Press, Chicago.

Bolonkin, A. A. (1969). Special Extrema in Optimal Control Problems, Engng. Cybern. No.2, 170-183.

Breakwell, J. V. (1965). A Doubly Singular Problem in Optimal Interplanetary Guidance, SIAM J. Control 3, 71-77.

Bryson, A. E. and Ho, Y. C. (1969). "Applied Optimal Control". Blaisdell, Waltham, Mass.

Bryson, A. E., Desai, M. N. and Hoffman, W. C. (1969). Energy-State Approximation in Performance Optimization of Supersonic Aircraft, J. Aircraft 6, 481-488.

Connor, M. A. (1974). Singular Control of Delay Systems, J. Opt. Th. Applic. 13, 538-544.

Dobell, A. R. and Ho, Y. C. (1967). Optimal Investment Policy: An Example of a Control Problem in Economic Theory, IEEE Trans. autom. Control AC-12, 4-14.

Forbes, G. F. (1950). The Trajectory of a Powered Rocket in Space, J. Br. Interplanet. Soc. 9, 75-79.

Fraeijs de Veubeke, B. (1965). Canonical Transformations and the Thrust-Coast-Thrust Optimal Transfer Problem, Astronautica Acta 11, 271-282.

Gabasov, R, (1968). Necessary Conditions for Optimality of Singular Control, Engng. Cybern. No.5, 28-37.

Gabasov, R. (1969). On the Theory of Necessary Optimality Conditions Governing Special Controls, Sov. Phys.-Dokl. 13, 1094-1095.

Gabasov, R. and Kirillova, F. M. (1972). High Order Necessary Conditions for Optimality, SIAM J. Control 10, 127-168.

Gabasov, R., Kirillova, F. M. and Strochko, V. A. (1971). Conditions for High-Order Optimality (Review), Automn remote Control 32, 689-704, 857-875, 1013-1040.

Gelfand, I. M. and Fomin, S. V. (1963). "Calculus of Variations". Prentice-Hall, Englewood Cliffs, N.J.

Goh, B. S. (1966a). The Second Variation for the Singular Bolza Problem, SIAM J. Control 4, 309-325.

Goh. B. S. (1966b). Necessary Conditions for Singular Extremals Involving Multiple Control Variables, SIAM J. Control 4, 716-731.

Goh, B. S. (1967). Optimal Singular Control for Multi-Input Linear Systems, J. math. Analysis Applic. 20, 534-539.

Goh, B. S. (1969/70). Optimal Control of a Fish Resource, Malay. Scientist 5, 65-70.

Goh, B. S. (1970). A Theory of the Second Variation in Optimal Control, unpublished report, Division of Applied Mechanics, Univ. California, Berkeley.

Goh, B. S. (1973). Compact Forms of the Generalized Legendre Conditions and the Derivation of Singular Extremals, Proc. 6th Hawaii International Conference on System Sciences, 115-117.

Graham, J. W. and D'Souza, A. F. (1970). Singular Optimal Control of Discrete-Time Systems, Proc. 11th JACC, 320-328.

Gurman, V. I. (1967). Method of Multiple Maxima and the Conditions of Relative Optimality of Degenerate Regimes, Automn remote Control 28, 1845-1852.

Haynes, G. W. (1966). On the Optimality of a Totally Singular Vector Control: An Extension of the Green's Theorem Approach to Higher Dimensions, SIAM J. Control 4, 662-677.

Hermes, H. (1964). Controllability and the Singular Problem, SIAM J. Control 2, 241-260.

Hermes, H. and Haynes, G. W. (1963). On the Nonlinear Control Problem with Control Appearing Linearly, SIAM J. Control 1, 85-108.

Jacobson, D. H. (1969). A New Necessary Condition of Optimality for Singular Control Problems, SIAM J. Control 7, 578-595.

Jacobson, D. H. (1970a). On Conditions of Optimality for Singular Control Problems, IEEE Trans. autom. Control AC-15, 109-110.

Jacobson, D. H. (1970b). Sufficient Conditions for Nonnegativity of the Second Variation in Singular and Nonsingular Control Problems, SIAM J. Control 8, 403-423.

Jacobson, D. H. (1970c). New Conditions for Boundedness of the Solution of a Matrix Riccati Differential Equation, J. Differential Equations 8, 258-263.

Jacobson, D. H. (1971a). A General Sufficiency Theorem for the Second Variation, J. math. Analysis Applic. 34, 578-589.

Jacobson, D. H. (1971b). Totally Singular Quadratic Minimization Problems, IEEE Trans. autom. Control AC-16, 651-658.

Jacobson, D. H. (1972). On Singular Arcs and Surfaces in a Class of Quadratic Minimization Problems, J. math. Analysis Applic. 37, 185-201.

Jacobson, D. H. and Speyer, J. L. (1971). Necessary and Sufficient Conditions for Optimality for Singular Control Problems: A Limit Approach, J. math. Analysis Applic. 34, 239-266.

Jacobson, D. H., Gershwin, S. B. and Lele, M. M. (1970). Computation of Optimal Singular Controls, IEEE Trans. autom. Control AC-15, 67-73.

Johansen, D. E. (1966). Convergence Properties of the Method of Gradients, in "Advances in Control Systems" (C. T. Leondes, ed) Vol.4, pp.279-316. Academic Press, New York and London.

Johnson, C. D. (1965). Singular Solutions in Optimal Control Problems, in "Advances in Control Systems" (C. T. Leondes, ed) Vol.2, pp.209-267. Academic Press, New York and London.

Johnson, C. D. and Gibson, J. E. (1963). Singular Solutions in Problems of Optimal Control, IEEE Trans. autom. Control AC-8, 4-15.

Keller, J. L. (1964). On Minimum Propellant Paths for Thrust Limited Rockets, Astronautica Acta 10, 262-269.

Kelley, H. J. (1963). Singular Extremals in Lawden's Problem of Optimal Rocket Flight, AIAA J. 1, 1578-1580.

Kelley, H. J. (1964a). A Second Variation Test for Singular Extremals, AIAA J. 2, 1380-1382.

Kelley, H. J. (1964b). A Transformation Approach to Singular Subarcs in Optimal Trajectory and Control Problems, SIAM J. Control 2, 234-240.

Kelley, H. J., Kopp, R. E. and Moyer, H. G. (1967). Singular Extremals, in "Topics in Optimization" (G. Leitmann, ed) pp.63-101. Academic Press, New York.

Kopp, R. E. and Moyer, H. G. (1965). Necessary Conditions for Singular Extremals, AIAA J. 3, 1439-1444.

Krener, A. J. (1973). The High Order Maximal Principle, in "Geometric Methods in Systems Theory" (D. Q. Mayne and R. W. Brockett, eds) pp.174-184. NATO Advanced Studies Institute Series: Mathematics and Physics.

Lawden, D. F. (1950). Note on a Paper by G. F. Forbes, J. Br. Interplanet. Soc. 9, 230-234.

Lawden, D. F.(1952). Inter-Orbital Transfer of a Rocket, Annual Report, Br. Interplanet. Soc., 321-333.

Lawden, D. F. (1961). Optimal Powered Arcs in an Inverse Square Law Field, J. Am. Rocket Soc. 31, 566-568.

Lawden, D. F. (1962). Optimal Intermediate-Thrust Arcs in a Gravitational Field, Astronautica Acta 8, 106-123.

Lawden, D. F. (1963). "Optimal Trajectories for Space Navigation", Butterworth, Washington, D.C.

Mancill, J. D. (1950). Identically Non-Regular Problems in the Calculus of Variations, Revista Matematica Y Fisica Teorica, Ser. A 7, 131-139.

Marchal, C. (1973). Chattering Arcs and Chattering Controls, J. Opt. Th. Applic. 11, 441-468.

Mayne, D. Q. (1970). Sufficient Conditions for Optimality for Singular Control Problems, Report 18/70, C.C.D., Imperial College, London.

Mayne, D. Q. (1973). Differential Dynamic Programming - A Unified Approach to the Optimization of Dynamic Systems, in "Advances in Control Systems (C. T. Leondes, ed) Vol.10, pp.179-254. Academic Press, New York and London.

McDanell, J. P. and Powers, W. F. (1970). New Jacobi-Type Necessary and Sufficient Conditions for Singular Optimization Problems, AIAA J. 8, 1416-1420.

McDanell, J. P. and Powers, W. F. (1971). Necessary Conditions for Joining Optimal Singular and Nonsingular Subarcs, SIAM J. Control 9, 161-173.

Miele, A. (1950-51). Problemi di Minimo Tempo nel Volo Non-Stazionario degli Aeroplani, Atti Accad. Sci. Torino 85, 41-52.

Miele, A. (1955a). General Solutions of Optimum Problems in Nonstationary Flight, NACA Tech. Memo. No.1388.

Miele, A. (1955b). Optimum Flight Paths of Turbojet Aircraft, NACA Tech. Memo. No.1389.

Miele, A. (1962). Extremization of Linear Integrals by Green's Theorem, in "Optimization Techniques" (G. Leitmann, ed) pp.69-98. Academic Press, New York.

Moore, J. B. (1969). A Note on a Singular Optimal Control Problem, Automatica 5, 857-858.

Moore, J. B. and Anderson, B. D. O. (1968). Extensions of Quadratic Minimization Theory. I Finite Time Results, Int. J. Control 7, 465-472.

Moyer, H. G. (1973). Sufficient Conditions for a Strong Minimum in Singular Control Problems, SIAM J. Control 11, 620-636.

Moylan, P. J. and Moore, J. B. (1971). Generalizations of Singular Optimal Control Theory, Automatica 7, 591-598.

Powers, W. F. and McDanell, J. P. (1971). Switching Conditions and a Synthesis Technique for the Singular Saturn Guidance Problem, J. Spacecraft Rockets 8, 1027-1032.

Robbins, H. M. (1965). Optimality of Intermediate-Thrust Arcs of Rocket Trajectories, AIAA J. 3, 1094-1098.

Robbins, H. M. (1967). A Generalized Legendre-Clebsch Condition for the Singular Cases of Optimal Control, IBM Jl Res. Dev. 3, 361-372.

Rohrer, R. A. (1968). Lumped Network Passivity Criteria, IEEE Trans. Circuit Theory CT-15, 24-30.

Rohrer, R. A. and Sobral, M. (1966). Optimal Singular Solutions for Linear Multi-Input Systems, Trans. ASME, J. Basic Eng. 88, 323-328.

Schultz, R. L. and Zagalsky, N. R. (1972). Aircraft Performance Optimization, J. Aircraft 9, 108-114.

Silverman, L. M. (1968). Synthesis of Impulse Response Matrices by Internally Stable and Passive Realizations, IEEE Trans. Circuit Theory CT-15, 238-245.

Sirisena, H. R. (1968). Optimal Control of Saturating Linear Plants for Quadratic Performance Indices, Int. J. Control 8, 65-87.

Snow, D. R. (1964). Singular Optimal Controls for a Class of Minimum Effort Problems, SIAM J. Control 2, 203-219.

Soliman, M. A. and Ray, W. H. (1972). On the Optimal Control of Systems having Pure Time Delays and Singular Arcs. I Some Necessary Conditions for Optimality, Int. J. Control 16, 963-976.

Speyer, J. L. (1973). On the Fuel Optimality of Cruise, J. Aircraft 10, 763-765.

Speyer, J. L. and Jacobson, D. H. (1971). Necessary and Sufficient Conditions for Optimality for Singular Control Problems: A Transformation Approach, J. math. Analysis Applic. 33, 163-187.

Steinberg, A. M. (1971). On Relaxed Control and Singular Solutions, J. Opt. Th. Applic. 8, 441-453.

Tait, K. S. (1965). Singular Problems in Optimal Control, Ph.D. dissertation, Harvard Univ., Cambridge, Mass.

Tarn, T. J., Rao, S. K. and Zaborszky, J. (1971). Singular Control of Linear-Discrete Systems, IEEE Trans. autom. Control AC-16, 401-410.

Vapnyarskiy, I. B. (1966). Solution of Some Optimal Control Problems, Engng. Cybern. No.4, 37-43.

Vapnyarskiy, I. B. (1967). An Existence Theorem for Optimal Control in the Bolza Problems, some of its Applications and the Necessary Conditions for the Optimality of Moving and Singular Systems, USSR Comp. Math., Math. Phys. 7, 22-54.

Vinter, R. B. (1974). Approximate Solution of a Class of Singular Control Problems, J. Opt. Th. Applic. 13, 461-483.

Wonham, W. M. and Johnson, C. D. (1964). Optimal Bang-Bang Control with Quadratic Performance Index, Trans. ASME, J. Basic Eng. 86, 107-115.

CHAPTER 2

Fundamental Concepts

2.1 Introduction

As has been described in Chapter 1 there was no standard mathematical procedure for analysing singular control problems before the early 1960's apart from Miele's 'Green's Theorem' approach. However, from about 1963 onwards researchers began to study the second variation of the general performance index of optimal control in an attempt to find new necessary conditions for a singular control to be optimal. The theory of the second variation of a functional had been well established in the classical calculus of variations and had figured prominently in the work carried out by Professor G. A. Bliss and his students at the University of Chicago during the first half of the present century. But although the general theory for the problem of Bolza reached a high degree of sophistication under the attention of the Chicago School no applications of this theory had been attempted. Indeed, in the preface to his book (Bliss, 1946) Professor Bliss makes an appeal for suitable examples to be listed which would illustrate the theory. This appeal has to a large extent been answered over the last thirty years by the enormous research effort engendered by control problems arising from the optimization of dynamical systems.

Until the early 1960's the theory of the second variation of a functional had rarely been applied to any practical problems. Even in the field of Mathematical Physics it had often been felt that the additional complexity of the second variation outweighed any possible benefit which might accrue from its use. However, with the advent of such singular problems as Lawden's intermediate-thrust arcs (Lawden, 1961, 1962, 1963) in Aerospace and Siebenthal and Aris's stirred tank reactor (Siebenthal and Aris, 1964) in Chemical Engineering it was clear that a satisfactory mathematical analysis of such problems lay in the theory of the second variation. This approach has been fully justified as will be seen in this book. Not only has the study of the second variation of a general performance index yielded necessary conditions for optimality of singular arcs but it has played a no less important part in the derivation of sufficient conditions for such arcs and in the production of algorithms for the numerical solution of non-singular problems.

This present chapter sets before the reader a few fundamental concepts necessary for an understanding of what is to follow. First, the general optimal control problem mentioned in Chapter 1 is reiterated and placed on a firm mathematical foundation. It should be emphasized here that we give a general statement of what is usually referred to as the optimal control problem. A singular problem is a special case of this general statement. Next, the first and second variations of the cost functional from the general optimal control problem are derived. The method of generation of these two variations follows closely that used by Bliss (1946) although, of course, his analysis does not include specific mention of control variables. Much of the notation used in this book coincides with that used by Bliss but where there are differences we have changed deliberately to be in keeping with that used in the modern control literature.

Having derived general expressions for both the first and second variations, shown that the first variation must be zero and the second variation non-negative for a minimizing arc, the final section of the present chapter formulates the general statement of a singular optimal control problem. The corresponding forms for the first and second variations in the singular case are stated. A number of examples are presented in this chapter to illustrate the many aspects of both variations.

2.2 The General Optimal Control Problem

Consider an n-dimensional state vector space X with time-varying elements $x = (x_1 x_2 \ldots x_n)^T$, $x_k = x_k(t)$, $k = 1,2,\ldots n$, and an m-dimensional control vector space V with elements $u = (u_1 u_2 \ldots u_m)^T$, $u_i = u_i(t)$, $i = 1,2,\ldots m$, $t_o \leq t \leq t_f$. The set V is defined by

$$V = \{u(t) : a_i \leq u_i \leq b_i, \quad i = 1,2,\ldots,m\} \tag{2.2.1}$$

where a_i, b_i can be known functions of time but are usually constants.

Since the major portion of this book will be discussing the case of singular control we shall assume, unless otherwise stated, that a control vector u belongs to the interior of space V so that $a_i < u_i < b_i$, $i = 1,2,\ldots,m$. Should some of the u_i's become equal to the corresponding bounds a_i or b_i then either the technique of Valentine (1937) may be employed or those variations $\beta_i(\cdot)$ (see below) of the control variables which attain their bounds can be put to zero. Furthermore, it may sometimes be convenient (in Section 3.2.2 for example) to update the control vector u to the status of a derivative and write

$$u = \dot{v}, \qquad v(t_o) = 0 \tag{2.2.2}$$

with $v(t_f)$ arbitrary. This transformation will be seen to bring the control problem more in line with the classical problem of Bolza.

Suppose the behaviour of a dynamical system is governed by differential equations

$$\dot{x} = f(x,u,t) \tag{2.2.3}$$

and boundary conditions

$$x(t_o) = x_o \tag{2.2.4}$$

$$\psi[x(t_f), t_f] = 0 \tag{2.2.5}$$

where t_o and x_o are specified and $\{t_o, x(t_o), t_f, x(t_f)\}$ belongs to S, a closed subset of R^{2n+2}. The terminal constraint function ψ is an s-dimensional column vector function of $x(t_f)$ and t_f. The final time t_f may or may not be specified.

We suppose further that the performance of the system is measured by a cost functional of the form

$$J = F[x(t_f), t_f] + \int_{t_o}^{t_f} L(x,u,t)dt. \tag{2.2.6}$$

The n-dimensional vector function f of eqn(2.2.3) and the scalar functions F and L are assumed to be at least twice continuously differentiable in each argument.

The general problem of optimal control is to find an element of U which minimizes the cost functional J of (2.2.6) subject to (2.2.3-5).

2.3 The First Variation of J

Define a one-parameter family of control vectors

$$u(\cdot, \varepsilon) \tag{2.3.1}$$

with the optimal vector given by $u(\cdot,0)$. The corresponding state vector

$$x(\cdot,\varepsilon) \tag{2.3.2}$$

will also be a function of time t ($t_o \leq t \leq t_f(\varepsilon)$) and parameter ε with $\varepsilon = 0$ along the optimal trajectory. In all cases $u(\cdot,\varepsilon)$ belongs to U, $x(\cdot,\varepsilon)$ belongs to some function space Y. Differentials of family (2.3.2) are

$$\begin{aligned} dt_f &= (dt_f/d\varepsilon)d\varepsilon, \qquad dx = \dot{x}\,dt + \delta x \\ d^2x &= \dot{x}\,d^2t + \ddot{x}\,dt^2 + 2\delta\dot{x}\,dt + \delta^2 x \end{aligned} \tag{2.3.3}$$

As in the classical calculus of variations (Bliss, 1946) the symbol δ denotes differentials only with respect to the parameter ε. We now introduce a set of state variations and final-time variations defined along the optimal trajectory as

$$\xi_f = (dt_f/d\varepsilon)_{\varepsilon=o}\,, \qquad \eta = (\partial x/\partial\varepsilon)_{\varepsilon=o} \tag{2.3.4}$$

in which case

$$dt_f = \xi_f d\varepsilon \text{ and } \delta x = \eta d\varepsilon.$$

Similarly, we can define a control variation along the optimal trajectory as

$$\beta = (\partial u/\partial \varepsilon)_{\varepsilon=o}. \tag{2.3.5}$$

From the system equations (2.2.3) it follows that the state variation η must satisfy the equation of variation

$$\dot{\eta} = f_x \eta + f_u \beta \tag{2.3.6}$$

where f_x, f_u are $n \times n$ and $n \times m$ matrices respectively.

Because of the boundary conditions (2.2.4-5) the set of variations ξ_f, η must satisfy end conditions of the form

$$\eta(t_o) = 0 . \tag{2.3.7}$$

$$\psi_{t_f}\xi_f + \psi_{x_f}(\dot{x}\xi_f + \eta)_{t=t_f} = 0 \tag{2.3.8}$$

where ψ_{t_f} is an s-dimensional vector and ψ_{x_f} an $s \times n$ matrix.

We now adjoin the system equations (2.2.3) and the terminal constraints (2.2.5) to the cost functional J of (2.2.6) by λ, an n-vector of Lagrange multiplier functions of time, and by ν, an s-dimensional constant vector of Lagrange multipliers respectively. The cost functional may then be written as

$$J = [F + \nu^T\psi]_{t=t_f} + \int_{t_o}^{t_f} \{H(x,u,\lambda,t) - \lambda^T\dot{x}\}dt \tag{2.3.9}$$

where $H(x,u,\lambda,t) = L(x,u,t) + \lambda^T f(x,u,t)$. (2.3.10)

When the vectors $u(\cdot,\varepsilon)$ and $x(\cdot,\varepsilon)$ of (2.3.1-2) are substituted into eqn(2.3.9) the cost functional J may be looked upon as a function of the single parameter ε and from $J(\varepsilon)$ one can easily calculate the first differential dJ. In fact,

$$dJ = [(F_t + \nu^T\psi_t + H - \lambda^T\dot{x})dt + (F_x + \nu^T\psi_x)dx]_{t=t_f} + \int_{t_o}^{t_f} \{H_x\delta x + H_u\delta u - \lambda^T\delta\dot{x}\}dt. \qquad (2.3.11)$$

By integrating the term in $\delta\dot{x}$ in the integrand of eqn(2.3.11) by parts, using $(2.3.3)_2$ to eliminate $\delta x(t_f)$ and noting that $\delta x(t_o) = 0$ since $x(t_o)$ is specified, we obtain

$$dJ = [(F_t + \nu^T\psi_t + H)dt + (F_x + \nu^T\psi_x - \lambda^T)dx]_{t=t_f} + \int_{t_o}^{t_f} \{(H_x + \dot{\lambda}^T)\delta x + H_u\delta u\}dt. \qquad (2.3.12)$$

On the optimal trajectory where the parameter ε is zero this differential takes the form $dJ = J_1(\xi_f,\eta,\beta)$. $d\varepsilon$. The second differential d^2J on the optimal trajectory can similarly be written as $d^2J = 2J_2(\xi_f,\eta,\beta)d\varepsilon^2$. A Taylor series for the function $J(\varepsilon)$ may then be written as

$$J(\varepsilon) = J(o) + \varepsilon J_1 + \varepsilon^2 J_2 + \ldots \qquad (2.3.13)$$

The function $J_1(\xi_f,\eta,\beta)$ is called the first variation of J on the optimal trajectory and from its definition and eqn(2.3.12) it is clear that

$$J_1(\xi_f,\eta,\beta) = [(F_t + \nu^T\psi_t + H)\xi + (F_x + \nu^T\psi_x - \lambda^T).$$

$$(\dot{x}\xi + \eta)]_{t=t_f} + \int_{t_o}^{t_f} \{(H_x + \dot{\lambda}^T)\eta + H_u\beta\}dt. \qquad (2.3.14)$$

Bearing in mind the assumption made in Section 2.2 that u belongs to the interior of V it follows from (2.3.13) that a necessary condition for $u(\cdot,0)$ to be a control vector which minimizes J is $J_1 = 0$. That is, a necessary condition for optimal control is that the first variation should vanish for all admissible variations. By choosing the adjoint vector λ and the vector ν so as to make the coefficients of η, $\eta(t_f)$ and ξ_f vanish in (2.3.14) we obtain the following results:

$$-\dot{\lambda}^T = H_x(\bar{x},\bar{u},\lambda,t) \qquad (2.3.15)$$

$$\lambda^T(t_f) = F_x(\bar{x}(t_f), t_f) + \nu^T\psi_{x_f} \qquad (2.3.16)$$

$$F_t(\overline{x}(t_f),\ t_f) + \nu^T\psi_{t_f} + H(\overline{x}(t_f),\ \overline{u}(t_f),\ \lambda,\ t_f) = 0. \tag{2.3.17}$$

The first variation of J then reduces to

$$J_1 = \int_{t_o}^{t_f} H_u\beta\ dt. \tag{2.3.18}$$

With the control variables away from any bounds the variation β in the integrand of (2.3.18) is arbitrary. Since J_1 is to vanish for all admissible variations β the fundamental lemma of the calculus of variations (Bliss, 1946) yields the condition

$$H_u = 0. \tag{2.3.19}$$

Of course, when u, a member of V, is allowed to attain its bounds we are led to Pontryagin's Minimum Principle, namely

$$\overline{u} = \arg\min_u H(\overline{x},u,\lambda,t). \tag{2.3.20}$$

Throughout the above discussion $\overline{x}(\cdot)$ and $\overline{u}(\cdot)$ denote the candidate state and control functions respectively.

A further first order condition is the necessary condition of Clebsch (Bliss, 1946). In the control formulation this condition may be written

$$\pi^T \begin{pmatrix} 0 & 0 \\ 0 & H_{uu} \end{pmatrix} \pi \geq 0 \tag{2.3.21}$$

for all (n+m)-vectors π satisfying the equation

$$(-I_n \ f_u)\pi = 0 \tag{2.3.22}$$

where I_n is the nth order identity matrix.

We now illustrate the use of necessary conditions (2.3.15), (2.3.19) to obtain a candidate arc for optimality by applying them to a rocket problem.

Example 2.1 The problem of finding the thrust direction programme necessary to maximize the range of a rocket with known propellant consumption is considered by Lawden (1963). The thrust direction is limited to lie in a vertical plane through the launching point. The acceleration due to gravity is assumed constant and flight takes place in vacuo over a flat earth. The rocket is launched with zero initial velocity at $t = 0$ and burn-out occurs at a known instant $t = T$. The vehicle continues under gravity along a ballistic trajectory until impact. The acceleration f caused by the motor thrust, essentially positive, is a given function of time.

With Ox and Oy horizontal and vertical axes through O and lying in the plane of flight, the equations of motion for this problem are

$$\begin{aligned} \dot{u} &= f\cos\theta \\ \dot{v} &= f\sin\theta - g \\ \dot{x} &= u \\ \dot{y} &= v \end{aligned} \tag{2.3.23}$$

where $f = cm/M$ and θ, the control variable, is the angle made by the thrust direction with Ox (Lawden, 1963).

The initial values of the state variables u, v (horizontal and vertical velocity) and x, y (horizontal and vertical displacement) are specified as is the time of flight T to burn-out. There are no end values specified at the final end-point except T. The boundary conditions for the problem are then

$$t_o = 0, \qquad u(0) = 0, \qquad v(0) = 0$$
$$x(0) = 0, \qquad y(0) = 0, \qquad t_f = T. \tag{2.3.24}$$

It is required to maximize the total range, which is a function of the values of the state variables at burn-out. This is equivalent to minimizing the cost function

$$J = F(x(t_f)) = -x_f - u_f[v_f + \sqrt{(v_f^2 + 2gy_f)}]/g \tag{2.3.25}$$

The Hamiltonian H of eqn(2.3.10) is

$$H = \lambda_u \, f\cos\theta + \lambda_v(f\sin\theta - g) + \lambda_x u + \lambda_y v. \tag{2.3.26}$$

Eqn(2.3.15) then yields

$$-\dot{\lambda}_u = \lambda_x, \quad -\dot{\lambda}_v = \lambda_y$$
$$\dot{\lambda}_x = 0, \quad \dot{\lambda}_y = 0. \tag{2.3.27}$$

Eqn(2.3.19) leads to the result

$$\tan\theta = \lambda_v/\lambda_u. \tag{2.3.28}$$

The end conditions given by eqn(2.3.16) are

$$\lambda_u(t_f) = -(v_f+r)/g, \quad \lambda_v(t_f) = -u_f(v_f+r)/gr,$$
$$\lambda_x(t_f) = -1, \quad \lambda_y(t_f) = -u_f/r \tag{2.3.29}$$

where $r = \surd(v_f^2 + 2gy_f)$. As in (Lawden, 1963) these results lead to

$$\tan\theta = u_f/r. \tag{2.3.30}$$

Pontryagin's Principle (2.3.20) or the Clebsch condition (2.3.21-22) are satisfied if θ takes the positive acute angle solution of eqn(2.3.30). The extremal is therefore a trajectory along which the thrust direction remains at a constant positive acute angle to Ox. Whether this extremal is an optimal trajectory is still to be decided and this will be the subject of a further example in the next section.

2.4 The Second Variation of J

In the previous section it was shown that the augmented cost functional J of (2.3.9) can be thought of as a function $J(\varepsilon)$ of the single parameter ε. Eqn(2.3.11) gives the first differential of J and is quoted again here for convenience:

$$dJ = [(F_t + \nu^T\psi_t + H - \lambda^T\dot{x})dt + (F_x + \nu^T\psi_x)dx]_{t=t_f}$$

$$+ \int_{t_o}^{t_f} \{H_x\delta x + H_u\delta u - \lambda^T\delta\dot{x}\}dt \qquad (2.4.1)$$

This formula is valid along all arcs $x(\cdot,\varepsilon)$, $u(\cdot,\varepsilon)$ belonging to the one-parameter families (2.3.1-2). In particular, we have seen that along the optimal trajectory (where ε=o) the first differential dJ vanishes.

From eqn(2.4.1) one may calculate the second differential of J as

$$d^2J = [(F_t + \nu^T\psi_t + H - \lambda^T\dot{x})d^2t + (F_x + \nu^T\psi_x)d^2x$$

$$+ (F_{tt} + \nu^T\psi_{tt} + \dot{H} - \lambda^T\ddot{x})dt^2$$

$$+ 2(F_{tx} + \nu^T\psi_{tx})dt\ dx + dx^T(F_{xx} + \nu^T\psi_{xx})dx$$

$$+ 2H_x\delta x dt + 2H_u\delta u dt - 2\lambda^T\delta\dot{x}dt]_{t=t_f}$$

$$+ \int_{t_o}^{t_f} \{H_x\delta^2x + H_u\delta^2u - \lambda^T\delta^2\dot{x} + \delta x^T H_{xx}\delta x$$

$$+ 2\delta u^T H_{ux}\delta x + \delta u^T H_{uu}\delta u\}dt. \qquad (2.4.2)$$

By expanding the term $\dot{H}$ = dH/dt into its constituent parts, integrating by parts the term in $\delta^2\dot{x}$, eliminating the δ-differentials outside the integral sign by means of eqns(2.3.3) and finally by using eqns(2.3.15, 19), we find that

$$d^2J = [(H + F_t + \nu^T\psi_t)d^2t + (F_x + \nu^T\psi_x - \lambda^T)d^2x]_{t_f}$$

$$+ [(F_{tt} + \nu^T\psi_{tt})dt^2 + 2(F_{tx} + \nu^T\psi_{tx})dt\ dx$$

$$+ dx^T(F_{xx} + \nu^T\psi_{xx})dx]_{t_f}$$

$$+ [(H_t - H_x\dot{x})dt^2 + 2H_x dxdt]_{t_f}$$

$$+ \int_{t_o}^{t_f} \{\delta x^T H_{xx}\delta x + 2\delta u^T H_{ux}\delta x + \delta u^T H_{uu}\delta u\}dt. \qquad (2.4.3)$$

The coefficients of the terms in d^2t_f and $d^2x(t_f)$ are zero because of the transversality conditions (2.3.16-17). The second differential thus reduces to

$$d^2J = \Phi_{t_f t_f} dt_f^{\ 2} + 2\Phi_{t_f x_f} dx_f dt_f + dx_f^{\ T}\ \Phi_{x_f x_f} dx_f$$

$$+ [(H_t - H_x\dot{x})dt^2 + 2H_x dx\ dt]_{t_f}$$

$$+ \int_{t_o}^{t_f} \{\delta x^T H_{xx}\delta x + 2\delta u^T H_{ux}\delta x + \delta u^T H_{uu}\delta u\}dt \qquad (2.4.4)$$

where $\Phi(x(t_f), t_f) = F + \nu^T\psi$ and $x_f \equiv x(t_f)$. As mentioned in Section 2.3 the second differential d^2J on the optimal trajectory may be written as $d^2J = 2J_2(\xi_f,\eta,\beta)d\varepsilon^2$ and from eqn(2.4.4) it is seen that

$$J_2 = \tfrac{1}{2}\Phi_{t_f t_f}\xi_f^{\,2} + \Phi_{t_f x_f}\xi_f(\dot{x}_f\xi_f + \eta(t_f))$$

$$+ \tfrac{1}{2}(\dot{x}_f\xi_f + \eta(t_f))^T\,\Phi_{x_f x_f}\,(\dot{x}_f\xi_f + \eta(t_f))$$

$$+ [\tfrac{1}{2}(H_t - H_x\dot{x})\xi^2 + H_x\xi(\dot{x}\xi + \eta)]_{t_f}$$

$$+ \int_{t_o}^{t_f}\{\tfrac{1}{2}\eta^T H_{xx}\eta + \beta^T H_{ux}\eta + \tfrac{1}{2}\beta^T H_{uu}\beta\}dt. \qquad (2.4.5)$$

This expression J_2 is the coefficient of ε^2 in the Taylor series (2.3.13) and is called the second variation of the cost functional J along the optimal trajectory. It follows that a necessary condition for $u(\cdot,0)$ to be a control vector which minimizes J is $d^2J/d\varepsilon^2 \geq 0$. That is to say the second variation J_2 must be non-negative.

A particular set of admissible variations satisfying eqns(2.3.6-8) is given by

$$\xi_f = 0, \qquad \eta(\cdot) \equiv 0, \quad \beta(\cdot) \equiv 0. \quad (2.4.6)$$

For this set of variations the second variation J_2 vanishes. Thus, if the candidate arc obtained from the vanishing of the first variation satisfies the condition $J_2 \geq 0$ then the set of variations (2.4.6) must minimize J_2. We are accordingly led to an auxiliary problem known as the accessory minimum problem. This is the problem of minimizing J_2 with respect to the set of variations ξ_f, $\eta(\cdot)$, $\beta(\cdot)$ satisfying the equations of variation (2.3.6–8).

A further necessary condition for a minimizing arc is known in the classical literature as the Jacobi condition. In the optimal control problem with second variation J_2 given above ($\xi_f = 0$) the Jacobi condition may be stated as follows (Bryson and Ho, 1969, but see also Chapter 4): an optimal trajectory contains no conjugate point between its end points. This will be the case if the matrix S remains finite for $t_o \leq t \leq t_f$ where S satisfies the matrix differential equation

$$-\dot{S} = H_{xx} + Sf_x + f_x^T S - (H_{xu} + Sf_u)H_{uu}^{-1}(H_{ux} + f_u^T S) \tag{2.4.7}$$

and end condition

$$S(t_f) = \Phi_{xx}[x(t_f), t_f]. \tag{2.4.8}$$

Example 2.2 (Bell, 1965) Consider again the problem of maximum range of a rocket vehicle discussed in the example of Section 2.3.

The equations of variation on the extremal are, from eqns(2.3.23),

$$\dot{\eta}_u = -\beta_\theta \, f\sin\theta \, , \qquad \dot{\eta}_v = \beta_\theta \, f\cos\theta$$
$$\dot{\eta}_x = \eta_u \, , \qquad \dot{\eta}_y = \eta_v . \tag{2.4.9}$$

The equations of variation on the extremal of the end conditions are, from eqns(2.3.24),

$$\xi_o = 0, \; \xi_f = 0, \; \eta_u(0) = 0, \; \eta_v(0) = 0,$$

$$\eta_x(0) = 0, \; \eta_y(0) = 0.$$

Using eqns(2.3.25-26, 29) the second variation (2.4.5) may be written as

$$J_2 = -\frac{1}{g}\left(1 + \frac{v_f}{r}\right)\eta_u(T)\,\eta_v(T) - \frac{1}{r}\,\eta_u(T)\,\eta_y(T)$$
$$+ \frac{u_f v_f}{r^3}\,\eta_v(T)\,\eta_y(T) - \frac{u_f y_f}{r^3}\,\eta_v^2(T) + \frac{u_f g}{2r^3}\,\eta_y^2(T)$$
$$- \tfrac{1}{2}\int_o^T f(t-k)\,\beta_\theta^2 \sec\theta \, dt$$

where $k = T + \frac{1}{g}(v_f + r)$. This variation has to be non-negative for the extremal found in Example 2.1 of Section 2.3 to be optimal. To demonstrate that it is indeed always non-negative we integrate eqns(2.4.9) with θ having the positive, acute angle solution of eqn(2.3.30). This gives

$$\eta_u(T) = -I_1 \sin\theta \quad , \quad \eta_v(T) = I_1 \cos\theta$$

$$\eta_x(T) = -I_2 \sin\theta \quad , \quad \eta_y(T) = I_2 \cos\theta$$

where $I_1 = \int_o^T f\beta_\theta dt$ and $I_2 = \int_o^T f(T-t)\beta_\theta dt$.

Substituting for $\eta_u(T)$, $\eta_v(T)$, $\eta_x(T)$ and $\eta_y(T)$ in the expression for J_2 we obtain

$$J_2 = \frac{gu_f}{r^3} \cos^2\theta[\frac{1}{g}(v_f + r)I_1 + I_2]^2$$

$$+ \sec\theta[\int_o^T f(T-t)\beta_\theta{}^2 dt + \frac{1}{g}(v_f + r)\int_o^T f\beta_\theta{}^2 dt].$$

Thus, $J_2 \geq 0$ and, moreover, $J_2 = 0$ if and only if $\beta_\theta \equiv 0$. The second variation is therefore always non-negative and the extremal found from the first variation satisfies the necessary condition associated with the second variation.

It is worth pointing out at this stage that the following notation for the second variation will normally be used throughout this book:

$$Q = H_{xx}, \quad C = H_{ux}, \quad R = H_{uu},$$
$$Q_f = \Phi_{x_f x_f}, \quad A = f_x, \quad B = f_u, \quad D = \psi_{x_f}. \tag{2.4.10}$$

Furthermore, the variations η in state and β in control will be denoted by x and u respectively. No confusion will result when the second variation is being considered as a new cost function $J[u(\cdot)]$ in the accessory minimum problem. The form of the second variation for a constrained optimal control problem, from eqns(2.4.5) and (2.3.6-8), may then be written as ($\xi_f = 0$)

$$J[(\cdot)] = \int_{t_o}^{t_f} (\tfrac{1}{2}x^T Qx + u^T Cx + \tfrac{1}{2}u^T Ru)\,dt + \tfrac{1}{2}x^T(t_f)Q_f x(t_f) \tag{2.4.11}$$

$$\text{subject to} \quad \dot{x} = Ax + Bu, \quad x(t_o) = 0. \tag{2.4.12}$$

$$\text{and} \quad Dx(t_f) = 0 \tag{2.4.13}$$

2.5 A Singular Control Problem

We consider here the class of control problems where the dynamical system is described by the ordinary differential equations

$$\dot{x} = f(x,u,t), \quad x(t_o) = x_o,$$

where $f(x,u,t) = f_1(x,t) + f_u(x,t)u$. (2.5.1)

The performance of the system is measured by the cost functional

$$J = \int_{t_o}^{t_f} L(x,t)dt + F[x(t_f), t_f] \tag{2.5.2}$$

and the terminal states must satisfy

$$\psi[x(t_f), t_f] = 0. \tag{2.5.3}$$

The final time t_f is assumed to be given explicitly. Thus, the Hamiltonian H for this problem formulation is linear in the control variables, and the problem turns out to be singular.

It is clear from eqn(2.4.5) that the second variation is

$$J_2 = \int_{t_o}^{t_f} (\tfrac{1}{2}\eta^T H_{xx}\eta + \beta^T H_{ux}\eta)dt + \tfrac{1}{2}\eta^T \Phi_{xx}\eta \Big|_{t_f} \tag{2.5.4}$$

subject to eqns(2.3.6-8). In terms of the notation mentioned at the end of Section 2.4 this second variation is

$$J[(\cdot)] = \int_{t_o}^{t_f} (\tfrac{1}{2}x^T Qx + u^T Cx)dt + \tfrac{1}{2}x^T(t_f)Q_f x(t_f) \tag{2.5.5}$$

subject to

$$\dot{x} = Ax + Bu, \quad x(t_o) = 0$$

$$Dx(t_f) = 0.$$

We conclude this section with a simple example:

Example 2.3 Consider the following scalar control problem: minimize

$$J = \tfrac{1}{2}\int_0^2 x^2 dt$$

subject to $\dot{x} = u, \quad x(0) = 1, \quad |u| \leq 1.$

This problem is linear in u with the Hamiltonian

$$H = \tfrac{1}{2}x^2 + \lambda u$$

where $\dot{\lambda} = -x.$

A singular arc is one along which

$$H_u = \lambda = 0$$

for a finite interval of time. During this interval we have

$$\dot{H}_u = \dot{\lambda} = 0$$

which implies $x = 0.$

In this case x vanishes identically and so does u. The arc in (x,t)-space along which u is zero is thus a singular arc.

References

Bell, D. J. (1965). Optimal Trajectories and the Accessory Minimum Problem, Aeronaut. Q. 16, 205-220.

Bliss, G. A. (1946). "Lectures on the Calculus of Variations". Univ. Chicago Press, Chicago.

Bryson, A. E. and Ho, Y. C. (1969). "Applied Optimal Control". Blaisdell, Waltham, Mass.

Lawden, D. F. (1961). Optimal Powered Arcs in an Inverse Square Law Field, J. Am. Rocket Soc. 31, 566-568.

Lawden, D. F. (1962). Optimal Intermediate-Thrust Arcs in a Gravitational Field, Astronautica Acta 8, 106-123.

Lawden, D. F. (1963). "Optimal Trajectories for Space Navigation", Butterworth, Washington, D.C.

Siebenthal, C. D. and Aris, R. (1964). Studies in Optimisation - VI. The Application of Pontryagin's Methods to the Control of a Stirred Reactor, Chem. Engng. Sci. 19, 729-746.

Valentine, F. A. (1937). The Problem of Lagrange with Differential Inequalities as Added Side Conditions, in "Contributions to the Theory of Calculus of Variations (1933-1937)" pp 403-447. Univ. Chicago Press, Chicago.

CHAPTER 3

Necessary Conditions for Singular Optimal Control

3.1 Introduction

It has already been stated in the previous two chapters that problems involving singular controls do arise in the optimization of engineering, economic and ecological systems. Because of this it is important that such problems be amenable to mathematical analysis in the same way as nonsingular problems are investigated. However, since Pontryagin's Principle does not yield any information directly on singular controls the first task for researchers was to discover new necessary conditions for optimality in the singular case.

This chapter describes some of the methods used to derive the two necessary conditions known as the generalized Legendre-Clebsch and Jacobson conditions. One approach for obtaining the former condition was based on a transformation of the singular problem to a nonsingular one (Kelley, 1964b). This transformation initially was applicable to problems with scalar control. It allows analysis of singular problems in a reduced state space but has the disadvantage of requiring the solution of a system of nonlinear differential equations in closed form. The generalized Legendre-Clebsch condition for scalar control was also obtained by using special control variations (Kelley, 1964a; Kopp and Moyer, 1965; Robbins, 1965). This method is described in Section 3.2.1 of the present chapter. The same condition but for vector control has been deduced by using the transformations of Goh (1966a, 1966b) and Kelley (Speyer and Jacobson, 1971). Section 3.2.2 of this chapter describes the transformation of Goh.

By using a particularly simple special control variation Jacobson deduced his necessary condition. In Section 3.3 Jacobson's condition is derived and shown to be a different necessary condition to the generalized Legendre-Clebsch; together they are insufficient for optimality.

Again a number of examples are presented in this chapter both to illustrate the theory and to draw attention to certain practical problems which give rise to singular control. Much of the work presented in this chapter plays an important part in later chapters where necessary and sufficient conditions are discussed. By using the necessary and sufficient conditions for the non-negativity of the second variation the two necessary conditions mentioned above are again derived in Chapter 4 for the totally singular case.

3.2 The Generalized Legendre-Clebsch Condition

3.2.1 Special Control Variations

Two different methods were used by those research workers who first derived a new necessary condition for singular optimal control problems with scalar control. This condition has since become known as the generalized Legendre-Clebsch condition due to its similarity in form to the classical conditions of Legendre and Clebsch. The two methods were due to Kelley (1964b) using a transformation to a state space of reduced dimension and to Robbins (1965), Kelley (1964a), Kopp and Moyer (1965) in which special control variations were employed. In the present section we give the derivation due to Kopp and Moyer. Both methods are discussed in detail by Kelley et al. (1967).

The method adopted to find new necessary conditions in the singular case is to generate special control variations and then to use the condition that the associated second variation must be non-negative. It is a technique that has been used to good effect in the classical literature on the nonsingular problem. The special control variations so generated must of course give rise to state variations which satisfy both the equations of variation (2.3.6) and the end conditions (2.3.7-8).

In the following derivation of the generalized Legendre-Clebsch condition a sequence of special control variations will be constructed which in turn will generate a sequence of necessary conditions. Should the first condition of this sequence be trivially satisfied then the second condition is tested and so on until new information is obtained.

Theorem 3.1. A necessary condition for the scalar control u in the control problem (2.5.1-3) to be optimal is that

$$\frac{\partial}{\partial u}\frac{d^2}{dt^2}H_u = \frac{\partial}{\partial u}\frac{d^2}{dt^2}\lambda^T B \leq 0$$

for all t in $[t_o, t_f]$.

Proof: A special scalar control variation, denoted by $\phi_o^1(\tau,\varepsilon)$ is shown in Fig.1.

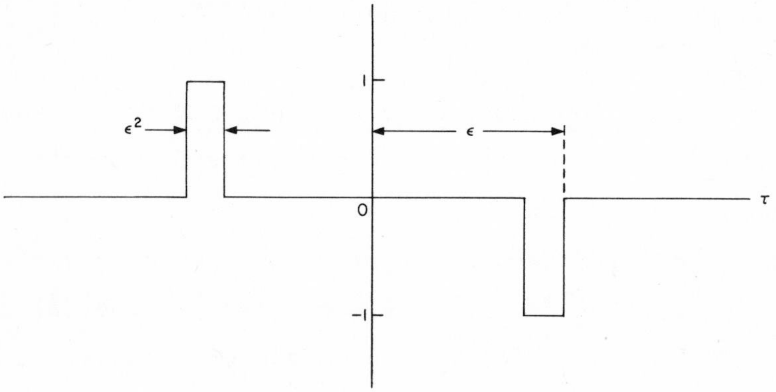

Fig. 1. Control variation $\phi_o^1(\tau,\varepsilon)$

The centre of this double pulse variation is at any point t_2 in (t_o, t_f). Introduce the fictitious time $\tau = t - t_2$ so that the variation is symmetrical about the origin $\tau = 0$ as shown. The interval of the double pulse is 2ε and the duration of each pulse is ε^2. Successive integrations of ϕ_o^1 with respect to t are designated $\phi_\nu^1(\tau,\varepsilon)$,

$$\text{i.e.} \qquad \frac{d^{\nu}\phi_{\nu}{}^{1}(\tau,\varepsilon)}{d\tau^{\nu}} = \phi_{o}{}^{1}(\tau,\varepsilon).$$

With the control variation $u = \phi_o{}^1(\tau,\varepsilon)$ the equations of variation (2.4.12) become

$$\dot{x} = Ax + B\phi_o{}^1(\tau,\varepsilon), \qquad x(t_o) = 0 \tag{3.2.1}$$

The end conditions at t_f (2.4.13) will be ignored for the time being. Assuming the necessary smoothness properties of the coefficients of (3.2.1) it can be demonstrated by direct substitution that a solution of (3.2.1) is given by

$$x = B\phi_1{}^1 + A_2\phi_2{}^1 + \theta^1 \tag{3.2.2}$$

where
$$A_2 = AB - \dot{B} \tag{3.2.3}$$

and
$$\dot{\theta}^1 = A\theta^1 + (AA_2 - \dot{A}_2)\phi_2{}^1, \qquad \theta^1(t_o) = 0 \tag{3.2.4}$$

With this expression for x substituted into the second variation (2.5.5) we obtain

$$J[(\cdot)] = \gamma[\xi_f, x(t_f)] + \int_{-\varepsilon}^{\varepsilon}\{\phi_o{}^1C(B\phi_1{}^1 + A_2\phi_2{}^1 + \theta^1)$$

$$+ \tfrac{1}{2}(\phi_1{}^1B^T + \phi_2{}^1A_2{}^T + \theta^{1T})Q(\phi_1{}^1B + \phi_2{}^1A_2 + \theta^1)\}d\tau$$

$$+ \text{contribution over } [\varepsilon, t_f - t_2]. \tag{3.2.5}$$

Integrating by parts the terms $\phi_o{}^1\phi_1{}^1CB$ and $\phi_o{}^1\phi_2{}^1CA_2$ and using the definitions of $\phi_1{}^1$ and $\phi_2{}^1$ the second variation may be expanded as a power series in ε, assuming the Taylor series in ε has a nonzero interval of convergence. Retaining the terms in $(\phi_1{}^1)^2$ which are the dominant terms in ε, we have

$$J = \left[-\frac{d}{dt}(CB) - 2C(AB-\dot{B}) + B^TQB\right]_{\tau=o} \varepsilon^5 + O(\varepsilon^6) \qquad (3.2.6)$$

Here we have used the fact that from equation (3.2.2)

$$x(t_f) = 2A_2(t_f)\varepsilon^3 + \theta^1(t_f) + O(\varepsilon^4)$$

and from equation (3.2.4) $\theta^1(t)$ is or order ε^3. It follows that γ, the quadratic form in $x(t_f)$ with $\xi_f = 0$, and the contribution to J over the time interval $[t_2 + \varepsilon, t_f]$ are of order ε^6. However, the state variations x of equation (3.2.2) corresponding to the special control variation $\phi_o{}^1(\tau,\varepsilon)$ will not in general satisfy the terminal constraints (2.4.13); they are of order ε^3. The question thus arises as to the admissibility of the variation $\phi_o{}^1(\tau,\varepsilon)$. However, since $x(t_f)$ is of order ε^3 we may alter the special control variation over the interval $[t_2 + \varepsilon, t_f]$ by the addition of an appropriate function of time which need only be of order ε^3. This additional function is chosen to ensure the satisfaction of the terminal constraints (2.4.13).

The resulting changes in the second variation (3.2.5) are of order ε^6 so that the dominant terms of that variation, stated in (3.2.6), are unchanged. The existence of such control correction functions is ensured by our assumption of normality.

Remembering that $\tau = 0$ is any interior point on the singular arc, a necessary condition for that arc to be minimizing is thus, from (3.2.6),

$$\frac{d}{dt}(CB) + 2C(AB - \dot{B}) - B^T QB \leq 0 \qquad (3.2.7)$$

This is identical to the result of Kelley (1964a) and was recognized by A. E. Bryson, Jr. as being equivalent to the compact form

$$\frac{\partial}{\partial u}\frac{d^2}{dt^2} H_u \leq 0 \qquad \text{for all } t \text{ in } [t_o, t_f]. \quad (3.2.8)$$

This condition is the first of the sequence of conditions which make up the generalized Legendre-Clebsch condition.

Should this condition (3.2.8) be trivially satisfied (equality) then no further information with which to test the singular arc has been obtained. In this case a second scalar control variation $\phi_o{}^2(\tau, \varepsilon)$ is constructed and is shown in Fig.2.

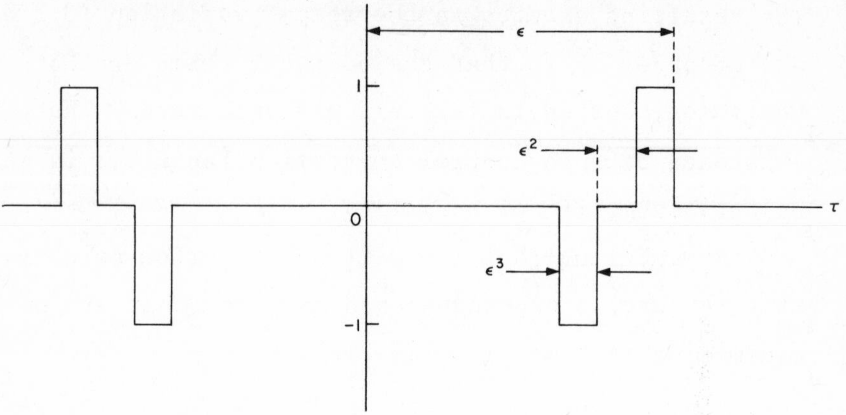

Fig. 2. Control variation $\phi_o^{\,2}(\tau,\varepsilon)$

Successive integrations of $\phi_o^{\,2}$ with respect to time are denoted by $\phi_\nu^{\,2}(\tau,\varepsilon)$,

$$\text{i.e.} \qquad \frac{d^\nu \phi_\nu^{\,2}(\tau,\varepsilon)}{d\tau^\nu} = \phi_o^{\,2}(\tau,\varepsilon).$$

Proceeding exactly as in the proof of Theorem 3.1, ignoring terminal constraints and with

$$x = B\phi_1^{\,2} + \sum_{i=2}^{4} A_i\phi_i^{\,2} + \theta^2,$$

$$A_2 = AB - \dot{B}, \quad A_3 = AA_2 - \dot{A}_2, \quad A_4 = AA_3 - \dot{A}_3,$$

$$\dot{\theta}^2 = A\theta^2 + (AA_4 - \dot{A}_4)\phi_4^{\,2},$$

one is led to the necessary condition

$$\frac{\partial}{\partial u}\frac{d^4}{dt^4} H_u \geq 0 \qquad \text{for all t in } [t_o,t_f]. \qquad (3.2.9)$$

Details of the derivation of (3.2.9) are given by Kelley et al. (1967). Again, the satisfaction of terminal constraints may be accomplished with control correction functions which do not affect the dominant term of the second variation from which the condition (3.2.9) is obtained.

If the condition (3.2.9) is trivially satisfied then a third control variation $\phi_o{}^3(\tau,\varepsilon)$ is constructed and so on. Details of the general analysis are also given by Kelley et al. (1967). The general result may be written as

$$(-1)^q \frac{\partial}{\partial u}\frac{d^{2q}}{dt^{2q}} H_u \geq 0. \qquad (3.2.10)$$

It can be shown (Kelley et al. 1967) that the control u cannot appear in an odd time derivative H_u when u is a scalar.

Example 3.1. The problem of control of a stirred reactor as defined by Siebenthal and Aris (1964) is that of determining the heat removal rate u and the associated state variables x_1 and x_2 denoting extent of reaction and temperature respectively, which satisfy differential equations

$$\dot{x}_1 = -x_1 + R(x_1, x_2)$$

$$\dot{x}_2 = R(x_1, x_2) - ax_2 - (x_2 + b)u,$$

where a and b are constants and R represents the rate of reaction such that a certain performance index may be extremized. In (Siebenthal and Aris, 1964) it is the time taken for steady state to be achieved from a given initial state which is to be minimized. All variables are measured from the steady state.

The Hamiltonian for this problem is

$$H = \lambda_1(-x_1 + R) + \lambda_2(R - ax_2 - (x_2 + b)u). \quad (3.2.11)$$

The adjoint variables λ_1 and λ_2 are governed by the equations

$$\dot{\lambda}_1 = \lambda_1(1 - R_1) - \lambda_2 R_1$$

$$\dot{\lambda}_2 = -\lambda_1 R_2 + \lambda_2(a + u - R_2)$$

where $R_1 = \partial R/\partial x_1$ and $R_2 = \partial R/\partial x_2$.

Since the Hamiltonian is linear in the control u there is the possibility of optimal singular control when the expression $\lambda_2(x_2 + b)$ vanishes identically over a finite interval of time. Furthermore it is shown by Siebenthal and Aris (1964) that such a singular control is possible in a reversible, exothermic reaction provided that

$$\lambda_2 \equiv 0, \quad R_2 \equiv 0 \text{ and } \lambda_1 \neq 0. \qquad (3.2.12)$$

We shall now apply the generalized Legendre-Clebsch condition to this singular arc (Bell, 1969).

From (3.2.11) we see that

$$H_u = -\lambda_2(x_2+b)$$

$$\frac{d}{dt} H_u = \lambda_1(x_2+b)R_2 - \lambda_2(ab-x_2R_2-bR_2+R)$$

and
$$\frac{d^2}{dt^2} H_u = \lambda_1 R_2(x_2+b) + (\lambda_1+\lambda_2)(x_2+b)[R_{21}(R-x_1) - R_1R_2 + R_{22}\{R-ax_2 - (x_2+b)u\}] + \lambda_1 R_2[R-ax_2 - (x_2+b)u] - \{ab-R_2(x_2+b) + R\}\{\lambda_2(a+u-R_2) - \lambda_1 R_2\} - \lambda_2 R_1(R-x_1)$$

where $R_{21} = \dfrac{\partial^2 R}{\partial x_2 \partial x_1}$ and $R_{22} = \dfrac{\partial^2 R}{\partial x_2{}^2}$.

Using the identities in (3.2.12) we see that

$$\frac{\partial}{\partial u} \frac{d^2}{dt^2} H_u = -\lambda_1(x_2+b)^2 R_{22}.$$

Since $x_2+b \neq 0$ (Siebenthal and Aris, 1964) the generalized Legendre-Clebsch condition (3.2.8) thus yields

$$\lambda_1 R_{22} \geq 0$$

along the singular arc. This is a new result, not obtainable from Pontryagin's Principle.

Example 3.2. Lawden (1963) has shown that a singular trajectory in the form of an intermediate-thrust arc satisfies Pontryagin's Principle and is a candidate for a minimum-fuel transfer orbit in space navigation when the final time is unspecified. This singular arc has become known as Lawden's spiral. The non-optimality of this singular arc has been demonstrated, for example by Kelley et al. (1967), using the generalized Legendre-Clebsch condition discussed above. This present example shows that Lawden's spiral is non-optimal by constructing special state variations which satisfy exactly the necessary end conditions on the variations (Bell, 1971).

The notation is that used by Lawden (1963). At time t the vehicle R of mass M has polar coordinates r,θ on the singular arc. At this instant the direction of the rocket's thrust makes an angle ϕ with the perpendicular to the radius vector drawn in the same sense as the motion. With O as pole and OX as initial line the angle made by the thrust direction and OX is ψ

(Fig.3).

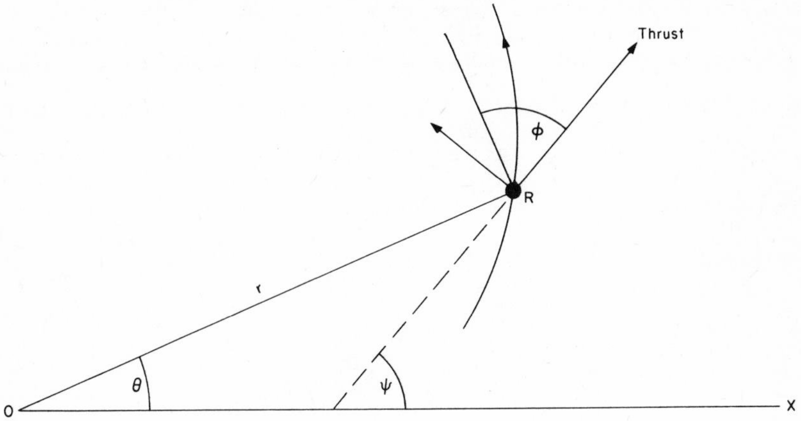

Fig. 3. Lawden's spiral

The vehicle's velocity components in the direction of and perpendicular to the thrust are (u,w). The equations of motion of the vehicle resolved in directions along and perpendicular to the thrust are

$$\dot{u} = \dot{\psi}w + \frac{cm}{M} - \frac{\gamma}{r^2}\sin\phi$$

$$\dot{w} = -\dot{\psi}u + \frac{\gamma}{r^2}\cos\phi$$

$$\dot{r} = u\sin\phi - w\cos\phi$$

$$r\dot{\theta} = u\cos\phi + w\sin\phi \qquad (3.2.13)$$

$$\dot{M} = -m$$

$$\theta = \psi + \phi - \frac{\pi}{2}$$

where c is the constant exhaust velocity of the rocket, m is the propellent mass flow rate and γ is the gravitational constant. The initial values of t, M, r, θ and the final values of r, θ are specified while the initial and final values of u, w and ϕ are given by the equations

$$
\begin{aligned}
u_o &= \dot{r}^o \sin\phi_o + r^o \dot{\theta}^o \cos\phi_o \\
w_o &= -\dot{r}^o \cos\phi_o + r^o \dot{\theta}^o \sin\phi_o \\
\psi_o + \phi_o &= \theta^o + \frac{\pi}{2} \\
u_f &= \dot{r}^f \sin\phi_f + r^f \dot{\theta}^f \cos\phi_f \\
w_f &= -\dot{r}^f \cos\phi_f + r^f \dot{\theta}^f \sin\phi_f \\
\psi_f + \phi_f &= \theta^f + \frac{\pi}{2} .
\end{aligned}
\tag{3.2.14}
$$

Suffices o and f denote as usual initial and final end points respectively. Superscripts o and f denote specified values at t_o and t_f. There are two control variables, namely ϕ and m.

It is required to determine the thrust magnitude and thrust direction along an intermediate-thrust arc which transfers the vehicle between the two fixed terminals with minimum consumption of fuel. A suitable cost function which is to be minimized is given by

$$J = c\ \ln(M_o/M_f).$$

It is clear from equations (3.2.13) that the Hamiltonian for this problem is linear in the mass flow rate m. This control variable m is singular. Results from the vanishing of the first variation J_1 are (Lawden, 1963)

$$w\dot{\psi} + \frac{\gamma}{r^2}\sin\phi = 0$$

$$w + r\dot{\psi}\sin\phi = A$$

$$\dot{\psi}^2 = \frac{\gamma}{r^3}(1 - 3\sin^2\phi)$$

$$\ddot{\psi} = -\frac{3\gamma}{r^3}\sin\phi\cos\phi$$

where A is a constant of integration. The equations of variation along the singular extremal are (2.3.6)

$$\begin{aligned}
\dot{\eta}_u &= \dot{\psi}\eta_w + w\dot{\eta}_\psi - \frac{d}{dt}\left(\frac{c}{M}\eta_M\right) + \frac{2\gamma}{r^3}\sin\phi\ \eta_r - \frac{\gamma}{r^2}\cos\phi\ \beta_\phi \\
\dot{\eta}_w &= -\dot{\psi}\eta_u - u\dot{\eta}_\psi - \frac{2\gamma}{r^3}\cos\phi\ \eta_r - \frac{\gamma}{r^2}\sin\phi\ \beta_\phi \\
\eta_u &= -w\beta_\phi + \sin\phi\ \dot{\eta}_r + \dot{\theta}\cos\phi\ \eta_r + r\cos\phi\ \dot{\eta}_\theta \\
\eta_w &= u\beta_\phi - \cos\phi\ \dot{\eta}_r + \dot{\theta}\sin\phi\ \eta_r + r\sin\phi\ \dot{\eta}_\theta \\
\dot{\eta}_M &= -\beta_m
\end{aligned} \qquad (3.2.15)$$

$$\eta_\psi + \beta_\phi = \eta_\theta .$$

The equations of variation on the extremal arc of the end conditions are (2.3.7-8)

$$\xi_o = \eta_M(0) = \eta_r(0) = \eta_\theta(0) = 0,$$

$$\eta_u(0) = -w_o\beta_\phi(0), \qquad \eta_w(0) = u_o\beta_\phi(0)$$

$$\eta_\psi(0) = -\beta_\phi(0) \tag{3.2.16}$$

$$\eta_r(t_f) = -\dot{r}_f\xi_f, \qquad \eta_\theta(t_f) = -\dot{\theta}_f\xi_f$$

$$\eta_u(t_f) = -\dot{u}_f\xi_f - w_f(\dot{\phi}_f\xi_f + \beta_\phi(t_f))$$

$$\eta_w(t_f) = -\dot{w}_f\xi_f + u_f(\dot{\phi}_f\xi_f + \beta_\phi(t_f)).$$

It can be shown (Bell, 1971) that the second variation reduces to

$$J_2 = \frac{2\gamma}{r^3}\dot{r}\sin\phi - \frac{\gamma}{r^2}(\dot{\theta} + \dot{\psi})\cos\phi\Big|_{t_f} \xi_f^2$$

$$+ \int_0^{t_f}\{f\eta_\psi^2 + \frac{6\gamma}{r^4}\sin\phi\ \eta_r^2 - \frac{6\gamma}{r^3}\cos\phi\ \eta_r\eta_\theta$$

$$- \frac{3\gamma}{r^2}\sin\phi\ \eta_\theta^2\}\ dt. \tag{3.2.17}$$

We now generate a special set of state variations η_r, η_θ satisfying the end conditions (3.2.16), to show that the second variation J_2 of (3.2.17) can assume a negative value, thus violating the necessary condition that it should be positive for all admissible variations. Choose

$$\begin{aligned} \eta_r &= p_1\ddot{G} + p_2\dot{G} - \dot{r}G \\ \eta_\theta &= q_1\ddot{G} + q_2\dot{G} - \dot{\theta}G \end{aligned} \tag{3.2.18}$$

where G(t) is an arbitrary function of time, except for certain end conditions specified below, and the functions $p_1(t)$, $p_2(t)$, $q_1(t)$, $q_2(t)$ are also arbitrary but finite functions of time subject only to the continuity conditions on the variations η_r, η_θ, η_u, η_w and η_M.

The end conditions (3.2.16) on the chosen variations η_r, η_θ(3.2.18) are satisfied provided function G(t) satisfies the boundary conditions

$$G(o) = 0, \qquad \dot{G}(o) = 0, \qquad \ddot{G}(o) = 0$$

$$G(t_f) = \xi_f, \qquad \dot{G}(t_f) = 0, \qquad \ddot{G}(t_f) = 0.$$

Differentiating equations (3.2.18) with respect to time, substituting into equations $(3.2.15)_{3-4}$ and employing equations $(3.2.13)_{3-4}$ yields expressions for the variations η_u and η_w. It turns out that these two latter variations satisfy end conditions (3.2.16) provided the function G(t) satisfies the further boundary conditions

$$G^{(iii)}(o) = 0 = G^{(iii)}(t_f).$$

Equation $(3.2.15)_2$ then yields an expression for $f\beta_\phi$ as a function of G and its derivatives up to and including $G^{(iv)}$.

Now we choose the functions $p_1(t)$, $p_2(t)$, $q_1(t)$, $q_2(t)$ so that the coefficients of the derivatives $G^{(iv)}$, $G^{(iii)}$ and $\ddot{G}$ vanish in the expression for $f(\beta_\phi - \eta_\theta)$. It can easily be deduced that this implies the following equations:

$$p_1\cos\phi = q_1 r\sin\phi$$

$$p_2\cos\phi - q_2 r\sin\phi = \frac{2\dot{r}\dot{\psi}}{\cos\phi} q_1 \tag{3.2.19}$$

$$2\dot{\psi}(p_2\sin\phi + q_2 r\cos\phi) = \frac{d}{dt}\left(\frac{2r\dot{\psi}}{\cos\phi} q_1\right) + w - \frac{6\gamma}{r^2}\sin\phi\, q_1.$$

From equations (3.2.19) the functions p_1, p_2 and q_2 can be found in terms of q_1. It is clear that the function $q_1(t)$ may be chosen arbitrarily provided it is non-zero. With these choices for the functions p_1, p_2 and q_2 the variation η_ψ is a function only of G(t) and its first derivative $\dot{G}(t)$. The second variation given by (3.2.17), after integration by parts and using end conditions on $\dot{G}(t)$, may then be written as

$$J_2 = \text{Terms in } \xi_f^{\,2} + \int_0^{t_f} \{ - \frac{3\gamma}{r^2} \frac{\sin\phi}{\cos^2\phi}(3-5\sin^2\phi) q_1^{\,2}\, \ddot{G}^2 + \text{terms in } \dot{G}^2 \text{ and } G^2 \} dt. \tag{3.2.20}$$

Now choose $G(t)$ such that

$$G(t) \equiv 0, \qquad 0 \leq t \leq t_1, \qquad 0 < t_1 < t_f$$

$$G(t) \equiv \xi_f, \qquad t_1 + \varepsilon \leq t \leq t_f$$

where ε is an arbitrarily small parameter. Furthermore, within the interval $[t_1, t_1 + \varepsilon]$ choose the fourth derivative $G^{(iv)}(t)$ to be of the form shown in Fig.4.

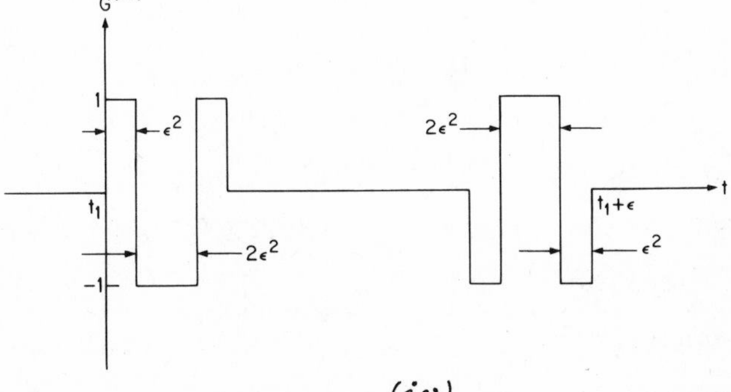

Fig. 4. The derivative $G^{(iv)}(t)$, $t_1 \leq t \leq t_1 + \varepsilon$.

The function G(t) and its first three derivatives are continuous.

It is easy to verify that G(t) satisfies all the required boundary conditions and furthermore,

$$G(t) \sim O(\varepsilon^8), \quad \dot{G}(t) \sim O(\varepsilon^6), \quad \ddot{G}(t) \sim O(\varepsilon^4),$$

$$G^{(iii)}(t) \sim O(\varepsilon^2).$$

In particular, $\xi_f \sim O(\varepsilon^8)$. The sign of the second variation is thus determined by the term in $\ddot{G}^2$ in (3.2.20); the dominant term, when ε is sufficiently small, being

$$-\int_{t_1}^{t_1+\varepsilon} \frac{3\gamma}{r^2} \frac{\sin\phi}{\cos^2\phi}(3-5\sin^2\phi)q_1{}^2\, \ddot{G}^2\, dt. \qquad (3.2.21)$$

Lawden (1963) has shown that the control variable ϕ must satisfy the inequalities

$$0 \leq \sin\phi \leq 1/\sqrt{3}.$$

Therefore, the term in (3.2.21) is negative and the second variation J_2 given by (3.2.20) can be made negative by choosing ε sufficiently small. This violates the necessary condition for optimal control that the second variation must be non-negative for a minimum value of J. Lawden's spiral cannot therefore form part of a minimum-fuel trajectory. Note that this result has been obtained with special variations which satisfy exactly the initial and terminal conditions (3.2.16).

3.2.2 A Transformation Approach

The generalized Legendre-Clebsch condition was derived in an alternative, indirect way by first transforming the singular problem into a nonsingular one and then applying the classical Legendre-Clebsch necessary condition (Kelley, 1964b; Kelley et al. 1967; Goh, 1966a, 1966b). This condition for the transformed problem is the generalized Legendre-Clebsch condition for the original singular problem. There have been two different transformations given in the literature. Kelley's transformation (Kelley, 1964b; Kelley et al., 1967; Speyer and Jacobson, 1971) reduces the dimension of the state space to n-p, where p is the order of the singular arc. The other transformation is due to Goh (1966a, 1966b) and is the one discussed in this section. It retains the full dimensionality of the original problem and is simpler in the case of vector controls (see also Chapter 6, Section 6.3).

As in Section 3.2.1 we start from the second variation for the singular arc

$$J_2 = \int_{t_o}^{t_f} \{\tfrac{1}{2}x^T Qx + u^T Cx\}dt + \tfrac{1}{2}x^T(t_f)Q_f\, x(t_f) \qquad (3.2.22)$$

subject to the equations of variation

$$\dot{x} = Ax + Bu, \qquad x(t_o) = 0 \tag{3.2.23}$$

and terminal conditions

$$Dx(t_f) = 0. \tag{3.2.24}$$

The basic idea behind Goh's transformation is to update the control vector to the status of a derivative of a new vector followed by the elimination of this derivative from both the second variation and equations of variation. A further updating of the new vector to a derivative may then yield a non-singular problem to which the classical Legendre-Clebsch condition may be applied. If this is not the case, then the procedure is repeated.

Define a vector v such that

$$\dot{v} = u \tag{3.2.25}$$

and, without loss of generality, set

$$v(t_o) = 0. \tag{3.2.26}$$

Furthermore, make the transformation from x to a new variable z:

$$z = x - Bv$$

in order to eliminate $\dot{v}$ from the equations of variation (3.2.23). These equations, together with the terminal conditions (3.2.24), become

$$\dot{z} = Az + (AB - \dot{B})v, \qquad z(t_o) = 0 \tag{3.2.27}$$

$$D[z(t_f) + Bv(t_f)] = 0. \tag{3.2.28}$$

The second variation becomes

$$J_2[v(\cdot)] = \int_{t_o}^{t_f} \{\tfrac{1}{2}z^TQz + v^TB^TQz + \tfrac{1}{2}v^TB^TQBv$$

$$+ \dot{v}^TC(z + Bv)\}dt$$

$$+ \tfrac{1}{2}[z(t_f) + Bv(t_f)]^TQ_f[z(t_f) + Bv(t_f)]. \tag{3.2.29}$$

Integrate by parts the term $\dot{v}^TCz$ by parts. This gives

$$J_2[v(\cdot)] = \int_{t_o}^{t_f} \{\tfrac{1}{2}z^TQz + v^T(B^TQ - CA - \dot{C})z$$

$$+ \tfrac{1}{2}v^T[B^TQB - 2C(AB - \dot{B})]v + \dot{v}CBv\}dt$$

$$+ \tfrac{1}{2}z^T(t_f)Q_fz(t_f) + v^T(t_f)[C + B^TQ_f]z(t_f)$$

$$+ \tfrac{1}{2}v^T(t_f)B^TQ_fBv(t_f). \tag{3.2.30}$$

But it can be shown that

$$B^TQB - 2C(AB - \dot{B}) - \frac{d}{dt}(CB) = -\frac{\partial}{\partial u}\ddot{H}_u$$

where, because of the identities (2.4.10),

$$H(x,u,\lambda,t) = \tfrac{1}{2}x^TQx + u^TCx + \lambda^T(Ax + Bu),$$

$$-\dot{\lambda} = H_x.$$

Therefore, the second variation given by (3.2.30) can be expressed as

$$J_2[v(\cdot)] = \int_{t_o}^{t_f} \{\tfrac{1}{2}z^TQz + v^T(B^TQ - CA - \dot{C})z$$

$$+ \tfrac{1}{2}v^T[-\frac{\partial}{\partial u}\ddot{H}_u]v + \dot{v}CBv + \tfrac{1}{2}v^T\frac{d}{dt}(CB)v\}dt$$

$$+ \tfrac{1}{2}z^T(t_f)Q_fz(t_f) + v^T(t_f)[C + B^TQ_f]z(t_f)$$

$$+ \tfrac{1}{2}v^T(t_f)B^TQ_fBv(t_f). \qquad (3.2.31)$$

Finally, integrating the last term in the integrand by parts,

$$J_2[v(\cdot)] = \int_{t_o}^{t_f} \{\tfrac{1}{2}z^TQz + v^T(B^TQ - CA - \dot{C})z$$

$$+ \tfrac{1}{2}v^T[-\frac{\partial}{\partial u}\ddot{H}_u]v + \tfrac{1}{2}\dot{v}^T[CB - (CB)^T]v\}dt$$

$$+ \tfrac{1}{2}z^T(t_f)Q_fz(t_f) + v^T(t_f)[C + B^TQ_f]z(t_f)$$

$$+ \tfrac{1}{2}v^T(t_f)[CB + B^TQ_fB]v(t_f). \qquad (3.2.32)$$

We pause at this point to deduce condition (1.5.3). Using an appropriate choice of control $v(\cdot)$ we can make the term in $\dot{v}$ dominate the other second-order quantities in (3.2.32). Since $CB - (CB)^T$ is skew-symmetric this means we can make J_2 either positive or negative. This violates the condition that J_2 must be of one sign for a minimizing or maximizing arc. We are thus led to the necessary condition that

$$CB = (CB)^T = B^T C^T \qquad \text{for all } t \text{ in } [t_o, t_f]. \quad (3.2.33)$$

This is a very powerful condition in the case of vector controls. It is easily verified (see proof of Lemma 4.2 in Chapter 4) that

$$CB - B^T C^T = \frac{\partial}{\partial u} \dot{H}_u .$$

Thus, (3.2.33) corresponds to (1.5.3) with $p = 1$. If u is a scalar then (3.2.33) is satisfied trivially.

Returning to equation (3.2.32) the term in $\dot{v}$ now vanishes by (3.2.33). The derivative $\dot{v}$ has thus been eliminated completely from the second variation and from the equations of variation. We may therefore update the control v to the status of a derivative, say $\dot{w}$. The second variation and differential constraints of the accessory minimum problem may now be written in the form

$$J_2[\dot{w}(\cdot)] = \int_{t_o}^{t_f} \{\tfrac{1}{2}\dot{\eta}^T R\dot{\eta} + \dot{\eta}^T L\eta + \tfrac{1}{2}\eta^T M\eta\}dt$$

$$+ \gamma[\dot{\eta}(t_f), \eta(t_f)], \qquad (3.2.34)$$

$$\phi\dot{\eta} + \theta\eta = 0 \qquad (3.2.35)$$

where

$$R = \begin{pmatrix} 0 & 0 \\ 0 & -\frac{\partial}{\partial u}\ddot{H}_u \end{pmatrix},$$

$$L = \begin{pmatrix} 0 & 0 \\ (B^T Q - CA - \dot{C}) & 0 \end{pmatrix}$$

$$M = \begin{pmatrix} Q & 0 \\ 0 & 0 \end{pmatrix}, \quad \phi = (I_n \quad (\dot{B} - AB)),$$

$$\theta = (-A \quad 0), \qquad \eta = (z^T \quad w^T)^T.$$

Applying the classical Clebsch condition (2.3.21-22) to the transformed accessory minimum problem (3.2.34-35) we obtain the following result: Along a minimizing arc

$$\pi^T R\pi \geq 0 \qquad (3.2.36)$$

for all (n+m)-vectors π satisfying

$$\phi\pi = 0. \qquad (3.2.37)$$

Taking $\pi_{n+1}, \pi_{n+2}, \ldots, \pi_{n+m}$ to be arbitrary, condition (3.2.36) subjected to (3.2.37) implies that

$$-\frac{\partial}{\partial u}\ddot{H}_u \geq 0 \qquad \text{for all t in } [t_o,t_f] \qquad (3.2.38)$$

which is (3.2.10) with q = 1. If (3.2.38) is met with equality for all t in $[t_o,t_f]$ then the functional $J_2[v(\cdot)]$ is totally singular and another transformation must be made. In this way the generalized necessary conditions (1.5.3-4) are obtained.

It should be noted that the derivative $\dot{w}(t_f)$ occurs in the terminal quadratic form γ of (3.2.34) and also in the terminal constraint (3.2.28) with $v = \dot{w}$. However, the class of admissible vectors w satisfying the differential constraints (3.2.35) is not reduced by restrictions placed upon the end values $\dot{w}(t_f)$. This is because any such restrictions can be met by infinitesimal adjustments over a sufficiently small neighbourhood

$[t_f-\varepsilon, t_f]$ and these adjustments will only affect the integral in J_2 infinitesimally. Consequently, the elements of the vector $\dot{w}(t_f)$ can be treated just as one would treat the parameter ξ_f (Goh, 1966a, 1966b; Speyer and Jacobson, 1971; McDanell and Powers, 1970).

Example 3.3 The problem of control of a particular aircraft model as defined by Schultz and Zagalsky (1972) is that of determining the minimum fuel required for a climb-cruise-descent trajectory between specified values of altitude, range and velocity. The control variables are thrust and flight path angle and the terminal time is not specified although the final result is valid also for fixed time. The system is governed by the differential equations and end conditions

$$\dot{E} = (T-D)V/M, \qquad E(t_o) = E^o, \qquad E(t_f) = E^f$$

$$\dot{h} = V\gamma, \qquad h(t_o) = h^o, \qquad h(t_f) = h^f$$

$$\dot{x} = V, \qquad x(t_o) = x^o, \qquad x(t_f) = x^f.$$

Superscripts o and f denote specified values at t_o and t_f respectively. E is the specific energy defined as

$$E = V^2/2 + gh,$$

h is the altitude, V the velocity, x the range, D(E,h) the aerodynamic drag and M the mass of the aircraft. The mass M will be assumed constant although the conclusions remain the same when this restriction is lifted. The control variables, thrust T and flight path angle γ, are bounded as

$$\gamma_{min} \leq \gamma \leq \gamma_{max}, \qquad T_{min}(E,h) \leq T \leq T_{max}(E,h).$$

It is required to minimize the fuel

$$J = \int_{t_o}^{t_f} \sigma(E,h)T \, dt$$

used during the manoeuvre, where $\sigma(E,h)$ is the fuel rate per unit thrust.

The Hamiltonian for this problem is

$$H = \sigma T + \lambda_1(T-D)V/M + \lambda_2 V\gamma + \lambda_3 V.$$

The adjoint variables λ_1, λ_2 and λ_3 are governed by the equations

$$\dot{\lambda}_1 = -\partial H/\partial E, \qquad \dot{\lambda}_2 = -\partial H/\partial h \quad \text{and} \quad \dot{\lambda}_3 = -\partial H/\partial x.$$

Since the Hamiltonian is autonomous and t_f is unspecified a first integral of the motion exists and is

$$H = 0 \qquad \text{for all } t \text{ in } [t_o, t_f].$$

The Hamiltonian is also linear in the control variables so that there is the possibility of optimal singular control when the expressions

$$\partial H/\partial T = \sigma + \lambda_1 V/M$$

and

$$\partial H/\partial \gamma = \lambda_2 V$$

vanish identically over a finite interval of time. This implies that

$$\lambda_1 = -\sigma M/V$$

$$\lambda_2 = 0.$$

It is shown by Schultz and Zagalsky (1972) that when the controls assume values which sustain the cruise portion of the trajectory

$$T = D, \qquad \gamma = 0.$$

Since the Hamiltonian is zero,

$$d(\partial H/\partial T)/dt = \partial(\sigma D/V)/\partial E = 0$$

$$d(\partial H/\partial \gamma)/dt = \partial(\sigma D/V)/\partial h = 0$$

which imply that the cruise is flown at the energy and altitude which satisfy $\min(\sigma D/V)$. The cruise condition is thus a doubly singular arc which satisfies the first order necessary conditions (Schultz and Zagalsky, 1972). We shall now apply the generalized Legendre-Clebsch condition to this singular arc (Speyer, 1973).

For the control vector $u^T = (T\ \gamma)$, eqn(3.2.33) (or equivalently eqn(1.5.3)) is not satisfied since

$$\begin{bmatrix} \frac{\partial}{\partial T}\frac{d}{dt}\frac{\partial H}{\partial T} & \frac{\partial}{\partial \gamma}\frac{d}{dt}\frac{\partial H}{\partial T} \\ \frac{\partial}{\partial T}\frac{d}{dt}\frac{\partial H}{\partial \gamma} & \frac{\partial}{\partial \gamma}\frac{d}{dt}\frac{\partial H}{\partial \gamma} \end{bmatrix} = \begin{bmatrix} 0 & (\frac{\partial\sigma}{\partial h} + \frac{\sigma g}{V^2})V \\ -(\frac{\partial\sigma}{\partial h} + \frac{\sigma g}{V^2})V & 0 \end{bmatrix}$$

and this is not the null matrix because the term $(\partial\sigma/\partial h + \sigma g/V^2)V$ in general is not zero. This result remains true if t_f is specified. The cruise condition will not therefore yield a minimizing singular arc. This result is contrary to the findings of Schultz and Zagalsky (1972).

3.3 Jacobson's Necessary Condition

A further necessary condition for non-negativity of $J_2[u(\cdot)]$ was discovered by Jacobson (1969, 1970b), different from the generalized Legendre-Clebsch condition. This new condition is stated and proved below for the free-endpoint problem, that is the matrix D in (3.2.24) is the null matrix. The appropriate condition for the constrained-endpoint problem is derived by Jacobson (1969). A slightly stronger version of Jacobson's condition for the free-endpoint problem was obtained by Gabasov (1968, 1969) although the conditions are identical for quadratic problems.

Theorem 3.2. A necessary condition for $J_2[u(\cdot)]$ to be non-negative for all $u(\cdot)$ belonging to U, for the case where D = 0, is that

$$CB + B^TWB \geq 0 \qquad \text{for all } t \text{ in } [t_o, t_f] \quad (3.3.1)$$

where

$$-\dot{W} = Q + A^TW + WA; \qquad W(t_f) = Q_f. \qquad (3.3.2)$$

Proof. Adjoin the equations of variation (3.2.23) to the expression J_2 of (3.2.22) by an adjoint vector of the form $\frac{1}{2}W^Tx$, where $W(\cdot)$ is an arbitrary n x n symmetric, continuously differentiable matrix function of time. Then

$$J_2[u(\cdot)] = \int_{t_o}^{t_f} \{\tfrac{1}{2}x^TQx + u^TCx + \tfrac{1}{2}x^TW(Ax+Bu-\dot{x})\}dt$$

$$+ \tfrac{1}{2}x^T(t_f)Q_f\, x(t_f). \qquad (3.3.3)$$

Integrate by parts the term in $\dot{x}$ to give

$$J_2[u(\cdot)] = \int_{t_o}^{t_f} \{\tfrac{1}{2}x^T(\dot{W} + Q + A^TW + WA)x$$

$$+ u^T(C + B^TW)x\}dt + \tfrac{1}{2}x^T(t_f)[Q_f - W(t_f)]x(t_f) \quad (3.3.4)$$

Now choose W to satisfy equations (3.3.2).

Then

$$J_2[u(\cdot)] = \int_{t_o}^{t_f} u^T(C + B^T W)x \, dt. \tag{3.3.5}$$

Choose a special variation u which is zero except during an interval $[t_1, t_1+\Delta]$, strictly contained in $[t_o, t_f]$, when it assumes a constant but arbitrary magnitude K. Then

$$J_2 = \int_{t_1}^{t_1+\Delta} K^T(C + B^T W)x \, dt \tag{3.3.6}$$

and the dominant term in the expansion of this integral for sufficiently small Δ is

$$\tfrac{1}{2}K^T(C + B^T W)BK\Delta^2\Big|_{t_1}. \tag{3.3.7}$$

A necessary condition for J_2 to be non-negative is therefore the condition (3.3.1) since t_1 can be chosen as any point in the interval $[t_o, t_f]$. (3.3.1) and (3.3.2) together form Jacobson's necessary condition.

The following example illustrates the non-equivalence of Jacobson's condition and the generalized Legendre-Clebsch condition. Furthermore, it demonstrates that these two necessary conditions are in general insufficient for optimality.

Example 3.4 (Jacobson, 1970a). Minimize

$$J = \tfrac{1}{2}\int_0^{3\pi/2} (-x_1^{\,2} + x_2^{\,2})\,dt \tag{3.3.8}$$

subject to

$$\begin{aligned} \dot{x}_1 &= x_2 , & x_1(o) &= 0 \\ \dot{x}_2 &= u , & x_2(o) &= 1. \end{aligned} \tag{3.3.9}$$

The Hamiltonian is given by

$$H = \tfrac{1}{2}(-x_1^{\,2} + x_2^{\,2}) + \lambda_1 x_2 + \lambda_2 u.$$

The adjoint equations for this problem are:

$$\dot{\lambda}_1 = x_1 , \qquad \lambda_1(3\pi/2) = 0$$

$$-\dot{\lambda}_2 = x_2 + \lambda_1, \qquad \lambda_2(3\pi/2) = 0.$$

It is easy to verify that

$$\begin{aligned} \bar{u}(t) &= -\sin t \\ \bar{x}_1(t) &= -\sin t \\ \bar{x}_2(t) &= \cos t \end{aligned} \tag{3.3.10}$$

is a singular solution satisfying Pontryagin's principle. For this solution, the magnitude of J is zero.

Here,

$$C = H_{ux} = (0 \;\; 0), \qquad B = f_u = (0 \;\; 1)^T$$

and it follows that

$$(C + B^T W)B = w_{22}$$

where

$$W = (w_{ij}), \qquad i,j = 1,2.$$

Using (3.3.2) it is easily verified that

$$w_{22}(\tau) = \tau - \frac{1}{3}\tau^3$$

where

$$\tau = 3\pi/2 - t,$$

i.e., reverse time. Hence, w_{22} is negative for $\tau > \sqrt{3}$ so that (3.3.1) ceases to be satisfied. The singular solution is then nonoptimal.

Also, for this problem

$$H_u = \lambda_2, \qquad \dot{H}_u = -x_2 - \lambda_1$$

$$\ddot{H}_u = -u - x_1$$

so that the generalized Legendre-Clebsch condition (3.2.38) is satisfied for all τ.

If (3.2.38) and (3.3.1) were together sufficient for optimality then the singular solution for $0 \leq \tau \leq \sqrt{3}$ would be optimal. That this is not the case can be demonstrated by transforming the original problem to a nonsingular one in the following way.

Write $z = x_1$ and $v = x_2$

so that the original problem becomes:

$$\text{Minimize } \tilde{J} = \tfrac{1}{2}\int_0^{3\pi/2} (-z^2 + v^2)\,dt \tag{3.3.11}$$

subject to

$$\dot{z} = v, \qquad z(o) = 0 \tag{3.3.12}$$

in which z is now a state variable and v a control variable. The problem is now nonsingular.

Clearly,

$$\bar{v}(t) = \cos t \tag{3.3.13}$$

satisfies Pontryagin's principle and $\tilde{J} = 0$ for this control. The strengthened Legendre-Clebsch condition ($H_{vv} > 0$) and the boundedness of the solution of the associated Riccati differential equation can be used to determine whether or not the control (3.3.13) is optimal since the problem is now nonsingular.

We have

$$H_{vv} = 1 > 0$$

and the Riccati equation is

$$-\frac{ds}{d\tau} = 1 + s^2, \qquad s = s(\tau), \qquad s(o) = 0.$$

The solution of this differential equation is

$$s(\tau) = -\tan \tau$$

which becomes unbounded at $\tau = \pi/2$. Therefore, the control (3.3.13) ceases to be optimal for $\tau > \pi/2$. This means that the singular solution (3.3.10) ceases to be optimal for $\tau > \pi/2$, since the two problems (3.3.8-9) and (3.3.11-12) are equivalent. But we have already shown from Jacobson's condition that (3.3.10) is nonoptimal for $\tau > \sqrt{3}$. Thus, we have the situation whereby the two necessary conditions are both satisfied for $\pi/2 < \tau < \sqrt{3}$ but the singular solution is nonoptimal. This shows that the satisfaction of both the generalized Legendre-Clebsch condition and the Jacobson condition is not sufficient for optimality in singular control problems.

In the derivations of the generalized Legendre-Clebsch condition it has been assumed that the problem is normal (Bliss, 1946). That is to say sufficient control variations exist to modify the special control variations in order to satisfy terminal constraints. Even in the absence of terminal constraints the assumption of normality is still necessary to insure the correct form of the test. In the proof of the Pontryagin Maximum Principle (Pontryagin et al., 1962) no assumption of normality is necessary since it is taken care of by the transversality condition on the adjoint variable and the additivity of first order variations. An extension of the Pontryagin Maximum Principle has been obtained by Krener (1973) and is called the High Order Maximal Principle. This extended principle includes not only the generalized Legendre-Clebsch and Jacobson conditions when they apply but also a number of other conditions which can be constructed for a specific problem without the assumption of normality.

References

Bell, D. J. (1969). Singular Extremals in the Control of a Stirred Reactor, Chem. Engng. Sci. 24, 521-525.

Bell, D. J. (1971). The Non-Optimality of Lawden's Spiral, Astronautica Acta 16, 317-324.

Bliss, G. A. (1946). "Lectures on the Calculus of Variations". Univ. Chicago Press, Chicago.

Gabasov, R. (1968). Necessary Conditions for Optimality of Singular Control, Engng. Cybern. No.5, 28-37.

Gabasov, R. (1969). On the Theory of Necessary Optimality Conditions Governing Special Controls, Sov. Phys.-Dokl. 13, 1094-1095.

Goh, B. S. (1966a). The Second Variation for the Singular Bolza Problem, SIAM J. Control 4, 309-325.

Goh, B. S. (1966b). Necessary Conditions for Singular Extremals Involving Multiple Control Variables, SIAM J. Control 4, 716-731.

Jacobson, D. H. (1969). A New Necessary Condition of Optimality for Singular Control Problems, SIAM J. Control 7, 578-595.

Jacobson, D. H. (1970a). On Conditions of Optimality for Singular Control Problems, IEEE Trans. autom. Control AC-15, 109-110.

Jacobson, D. H. (1970b). Sufficient Conditions for Nonnegativity of the Second Variation in Singular and Nonsingular Control Problems, SIAM J. Control 8, 403-423.

Kelley, H. J. (1964a). A Second Variation Test for Singular Extremals, AIAA J. 2, 1380-1382.

Kelley, H. J. (1964b). A Transformation Approach to Singular Subarcs in Optimal Trajectory and Control Problems, SIAM J. Control 2, 234-240.

Kelley, H. J., Kopp, R. E. and Moyer, H. G. (1967). Singular Extremals, in "Topics in Optimization" (G. Leitmann, ed) pp 63-101. Academic Press, New York.

Kopp, R. E. and Moyer, H. G. (1965). Necessary Conditions for Singular Extremals, AIAA J. 3, 1439-1444.

Krener, A. J. (1973). The High Order Maximal Principle, in "Geometric Methods in Systems Theory" (D. Q. Mayne and R. W. Brockett, eds) pp 174-184. NATO Advanced Studies Institute Series: Mathematics and Physics.

Lawden, D. F. (1963). "Optimal Trajectories for Space Navigation", Butterworth, Washington, D.C.

McDanell, J. P. and Powers, W. F. (1970). New Jacobi-Type Necessary and Sufficient Conditions for Singular Optimization Problems, AIAA J. 8, 1416-1420.

Pontryagin, L. S., Boltyanskii, V. G., Gamkrelidze, R. V. and Mishchenko, E. F. (1962). "The Mathematical Theory of Optimal Processes". Interscience, John Wiley & Sons, Inc., New York, London and Sydney.

Robbins, H. M. (1965). Optimality of Intermediate-Thrust Arcs of Rocket Trajectories, AIAA J. 3, 1094-1098.

Schultz, R. L. and Zagalsky, N. R. (1972). Aircraft Performance Optimization, J. Aircraft 9, 108-114.

Siebenthal, C. D. and Aris, R. (1964). Studies in Optimisation - VI. The Application of Pontryagin's Methods to the Control of a Stirred Reactor, Chem. Engng. Sci. 19, 729-746.

Speyer, J. L. (1973). On the Fuel Optimality of Cruise, J. Aircraft 10, 763-765.

Speyer, J. L. and Jacobson, D. H. (1971). Necessary and Sufficient Conditions for Optimality for Singular Control Problems; A Transformation Approach, J. math. Analysis Applic. 33, 163-187.

CHAPTER 4

Sufficient Conditions and Necessary and Sufficient Conditions for Non-Negativity of Nonsingular and Singular Second Variations

4.1 Introduction

As indicated in Chapter 2, non-negativity of the second variation is a (second-order) necessary condition for optimality in control problems. It is known, also, that a sufficient condition for a weak local minimum is that the second variation be strongly positive (Gelfand and Fomin, 1963). Unfortunately, it turns out that in singular control problems the second variation *cannot* be strongly positive (Tait, 1965; Johansen, 1966) and so different tests for sufficiency have to be devised. Despite this fact, the second variation continues to play an important role in second-order optimality tests and it is for this reason that this chapter explores the second variation fully.

In the case of nonsingular optimal control problems, where $H_{uu} > 0$ for all t in $[t_o, t_f]$, it is known that a sufficient condition for strong positivity of the second variation, and hence for a weak local minimum, is that the matrix Riccati differential equation associated with the second variation should have a solution for all t in $[t_o, t_f]$. Clearly, this condition is inapplicable in the singular case owing to the presence of H_{uu}^{-1} in the matrix Riccati equation. For a long time, therefore, it was felt that no Riccati-like condition existed for the singular case but this has turned out to be not true (Goh, 1970; Jacobson, 1970a, 1971a, 1971b; Jacobson and Speyer, 1971; McDanell and Powers, 1970; Speyer and Jacobson, 1971). The result is that sufficiency conditions for non-negativity of singular and non-singular second variations are rather closely related.

We begin by proving that a necessary and sufficient condition for strong positivity of the nonsingular second variation is that a solution exists to the well known matrix Riccati differential equation. The sufficiency part of this theorem is very well known and documented (Brockett, 1970) but though the necessity part is well known it is not, in our opinion, proved convincingly elsewhere except in certain special cases. Following this result, we then illustrate that the singular second variation cannot be strongly positive and confirm that the standard Riccati equation is of no direct use. Next, we provide a sufficient condition (Jacobson, 1971a) for the partially singular second variation which is in the form of a set of differential and algebraic inequalities. We show how, in the non-singular case, the standard Riccati equation is related to these conditions, and thereby implicitly establish also the necessity of these conditions. In the totally singular case we show that a new Riccati differential equation is related to the sufficiency conditions. We then turn our attention to necessary and sufficient conditions for non-negativity of the totally singular second variation and produce again a set of differential and algebraic inequalities. These conditions are slightly more abstract in form than the ones presented earlier and, because they are necessary as well as sufficient, all known necessary conditions can be deduced from these. Recent extensions of the conditions (Anderson, 1973) are mentioned. Finally, the role of the conditions in deducing sufficient conditions for a weak local minimum in singular control problems and in deducing existence conditions for a solution to the standard matrix Riccati equation, is investigated.

4.2 Preliminaries

The form of the second variation for an unconstrained optimal control problem is,+ from Chapter 2,

$$J[u(\cdot)] = \int_{t_o}^{t_f} (\tfrac{1}{2}x^T Qx + u^T Cx + \tfrac{1}{2}u^T Ru)\,dt + \tfrac{1}{2}x^T(t_f)Q_f x(t_f) \tag{4.2.1}$$

subject to

$$\dot{x} = Ax + Bu; \quad x(t_o) = 0 \tag{4.2.2}$$

+ We use here the notation $J[u(\cdot)]$ for the second variation. This is done for simplicity of presentation.

where x is in R^n, u is in R^m, Q,C,R,A,B are continuous matrix functions of time of appropriate dimensions, and where Q_f is in R^{nxn} and is constant. Without loss of generality, it is assumed that Q, R and Q_f are symmetric.

We now make three explicit assumptions; one on the class of allowed controls and the others on the nature of R and of (4.2.2).

Assumption 4.1 The class of controls U from which u(·) is drawn is the class of piecewise continuous m-vector functions of time on $[t_o,t_f]$.

Though a less restrictive class of controls could be assumed, the class of piecewise continuous functions is adequate for the problems treated in this book.

Assumption 4.2 A basic assumption that we make is that

$$R(t) \geq 0 \qquad \text{for all } t \text{ in } [t_o,t_f]. \qquad (4.2.3)$$

In fact this assumption is not at all restrictive since if R were not positive semi-definite the criterion (4.2.1) would, regardless of any other conditions, be able to take on arbitrarily large, negative, values. Assumption 4.2 is the Legendre-Clebsch necessary condition for optimality in the calculus of variations (Gelfand and Fomin, 1963).

Assumption 4.3 We assume that the dynamic system (4.2.2) is completely controllable on the interval

$[t_o, t']$, where $t_o < t' \leq t_f$. That is,

$$\int_{t_o}^{t'} \Phi(t',\sigma)B(\sigma)B^T(\sigma)\Phi^T(t',\sigma)\,d\sigma > 0 \qquad (4.2.4)$$

where

$$\frac{\partial}{\partial t}\Phi(t,\sigma) = A(t)\Phi(t,\sigma) \;;\; \Phi(\sigma,\sigma) = I \qquad (4.2.5)$$

for all t' in $(t_o, t_f]$.

Note that this assumption is used in only some of the results derived in this chapter.

As the main objective of this chapter is to develop conditions under which (4.2.1) is non-negative, positive definite, and strongly positive, we now define these terms precisely.

Definition 4.1 $J[u(\cdot)]$ is said to be non-negative if for each $u(\cdot)$ in U its value is non-negative.

Definition 4.2 $J[u(\cdot)]$ is said to be positive definite if for each $u(\cdot)$ in U, $u(\cdot) \neq \theta$ (null function), $J[u(\cdot)] > 0$.

Definition 4.3 $J[u(\cdot)]$ is said to be strongly positive if for each $u(\cdot)$ in U, and some $k > 0$,

$$J[u(\cdot)] \geq k\|u(\cdot)\|^2 \qquad (4.2.6)$$

where $\|u(\cdot)\|$ is some suitable norm defined on U.

4.3 The Nonsingular Case

Definition 4.4 We refer to (4.2.1) as nonsingular if

$$R(t) > 0, \quad \text{for all } t \text{ in } [t_o, t_f]. \tag{4.3.1}$$

In this case we have the following theorem.

Theorem 4.1 A necessary and sufficient condition for $J[u(\cdot)]$ to be strongly positive is that there exists for all t in $[t_o, t_f]$ a function $S(\cdot)$ which satisfies the Riccati equation

$$-\dot{S} = Q + SA + A^T S - (C + B^T S)^T R^{-1} (C + B^T S) \tag{4.3.2}$$

$$S(t_f) = Q_f. \tag{4.3.3}$$

In order to prove this theorem we require the following lemma.

Lemma 4.1 Suppose that $t_o < t' \leq t_f$ and that (4.2.2) is completely controllable on $[t_o, t']$. Then, there exists $u^*(t)$, $t_o \leq t \leq t'$, which transfers $x(t_o) = 0$ to any desired $x_{t'}$ at $t = t'$ such that

$$||x_{t'}|| = 1 \tag{4.3.4}$$

and

$$||u^*(\cdot)|| \leq \rho(t') \tag{4.3.5}$$

where $\rho(\cdot)$ is a continuous function defined on $(t_o, t']$.

Proof: Define

$$W(t_o,t') = \int_{t_o}^{t'} \Phi(t',t)B(t)B^T(t)\Phi^T(t',t)dt \qquad (4.3.6)$$

and

$$u^*(t)=B^T(t)\Phi^T(t',t)W^{-1}(t_o,t')x_{t'} \; . \qquad (4.3.7)$$

It is a trivial matter to verify that the control function $u^*(\cdot)$ transfers the state of system (4.2.2) from the origin at t_o to $x_{t'}$ at $t=t'$. Furthermore, since $\Phi(t'\cdot)$ and $W^{-1}(t_o,t')$ are continuous in t' and $||x_{t'}|| = 1$, it is clear that there is a continuous function $\rho(\cdot)$ such that

$$||u^*(\cdot)|| \leq \rho(t'). \qquad (4.3.8)$$

Proof of Theorem 4.1 First we prove that if $S(\cdot)$ satisfies (4.3.2) and (4.3.3) then $J[u(\cdot)]$ is positive definite.

Consider the identically zero quantity

$$\tfrac{1}{2}x^T(t_o)S(t_o)x(t_o)-\tfrac{1}{2}x^T(t_f)S(t_f)x(t_f)$$
$$+ \int_{t_o}^{t_f} \tfrac{1}{2}\frac{d}{dt}[x^T(t)S(t)x(t)]dt \qquad (4.3.9)$$

which, re-written using (4.3.2) and (4.3.3) is

$$\tfrac{1}{2}x^T(t_o)S(t_o)x(t_o)-\tfrac{1}{2}x^T(t_f)Q_f x(t_f)$$

$$+\int_{t_o}^{t_f}\{-\tfrac{1}{2}x^T[Q+SA+A^TS-(C+B^TS)^TR^{-1}(C+B^TS)]x$$

$$+\tfrac{1}{2}x^T(SA+A^TS)x+x^TSBu\}dt. \qquad (4.3.10)$$

Adjoining (4.3.10) to (4.2.1) gives

$$J[u(\cdot)] = \tfrac{1}{2}x^T(t_o)S(t_o)x(t_o)$$

$$+\int_{t_o}^{t_f}\tfrac{1}{2}[u+R^{-1}(C+B^TS)x]^TR[u+R^{-1}(C+B^TS)x]dt \qquad (4.3.11)$$

which takes on its minimum value of zero (because $x(t_o) = 0$) if and only if

$$u(t) = -R^{-1}(t)[C(t)+B^T(t)S(t)]x(t), \qquad (4.3.12)$$

$$t_o \le t \le t_f.$$

That is, if and only if,

$$u(\cdot) = \theta \text{ (null function)}. \qquad (4.3.13)$$

Hence, $J[u(\cdot)]$ is positive definite.

Now consider

$$J[u(\cdot),\varepsilon] = \int_{t_o}^{t_f} \{\tfrac{1}{2}x^TQx+u^TCx+ \frac{1}{2-\varepsilon}\, u^TRu\}dt + \tfrac{1}{2}x^T(t_f)Q_fx(t_f). \tag{4.3.14}$$

This functional is positive definite if $2-\varepsilon>0$ and if

$$-\dot{S}(t,\varepsilon)=Q+S(t,\varepsilon)A+A^TS(t,\varepsilon)-[C+B^TS(t,\varepsilon)]^T\frac{(2-\varepsilon)}{2}\, R^{-1} [C+B^TS(t,\varepsilon)] \tag{4.3.15}$$

$$S(t_f,\varepsilon) = Q_f \tag{4.3.16}$$

has a solution $S(t,\varepsilon)$ defined for all t in $[t_o,t_f]$.

Now, since Q,C,R,A,B are continuous in t and the right hand side of (4.3.15) is analytic in $S(t,\varepsilon)$ and ε, and since $S(\cdot,0)$exists, we have that $S(\cdot,\varepsilon)$ is a continuous function of ε at $\varepsilon=0$ (Coddington and Levinson, 1955). So, for ε sufficiently small, $S(t,\varepsilon)$ exists for all t in $[t_o,t_f]$. Therefore, for $\varepsilon<0$, and sufficiently small, $J[u(\cdot),\varepsilon]$ is positive definite.

Next, we note that

$$J[u(\cdot),\varepsilon] = J[u(\cdot)] + \frac{\varepsilon}{2(2-\varepsilon)}\int_{t_o}^{t_f} u^TRudt \geq 0 \tag{4.3.17}$$

so that

$$J[u(\cdot)] \geq -\frac{\varepsilon}{2(2-\varepsilon)}\int_{t_o}^{t_f} u^T R u dt \qquad (4.3.18)$$

and because of (4.3.1) we conclude from (4.3.18) that

$$J[u(\cdot)] \geq -\frac{\varepsilon k_1}{2(2-\varepsilon)}\int_{t_o}^{t_f} u^T u dt, \quad k_1 > 0 \qquad (4.3.19)$$

so that

$$J[u(\cdot)] \geq k||u(\cdot)||^2 \qquad (4.3.20)$$

where

$$k = -\frac{\varepsilon k_1}{2(2-\varepsilon)} > 0 \qquad (4.3.21)$$

and

$$||u(\cdot)||^2 = \int_{t_o}^{t_f} u^T u dt \qquad (4.3.22)$$

which implies that $J[u(\cdot)]$ is strongly positive.

Now we prove the converse; namely, that if $J[u(\cdot)]$ is strongly positive, then an $S(\cdot)$ exists which satisfies (4.3.2) and (4.3.3).

Define

$$J'[u(\cdot),x(t'),t'] = \int_{t'}^{t_f}[\tfrac{1}{2}x^T Q x + u^T C x + \tfrac{1}{2}u^T R u]dt$$
$$+ \tfrac{1}{2}x^T(t_f)Q_f x(t_f). \qquad (4.3.23)$$

Then, for t' sufficiently close to t_f we have by standard differential equation theory (Coddington and Levinson, 1955) that $S(t)$ exists, $t' \leq t \leq t_f$. Consequently, the minimum value of (4.3.23), denoted by $J[u^o(\cdot),x(t'),t']$, is given by (see (4.3.11) and (4.3.12))

$$J[u^o(\cdot),x(t'),t'] = \tfrac{1}{2}x^T(t')S(t')x(t') \qquad (4.3.24)$$

where

$$u^o(t) = -R^{-1}(t)[C(t)+B^T(t)S(t)]x(t), \qquad (4.3.25)$$

$$t' < t \leq t_f.$$

Now suppose that $S(t)$ ceases to exist at t_e where

$$t_o < t_e < t' \leq t_f . \qquad (4.3.26)$$

By Lemma 4.1, we can construct any desired $x(t')$, having $||x(t')|| = 1$, using some control function $u^*(t)$, $t_o \leq t \leq t'$ which satisfies

$$||u^*(\cdot)|| \leq \rho(t') < \infty, \quad t_o < t' \leq t_f. \qquad (4.3.27)$$

Now,

$$J[u(\cdot)] = \int_{t_o}^{t'} [\tfrac{1}{2}x^TQx+u^{*T}Cx+\tfrac{1}{2}u^{*T}Ru^*]dt$$
$$+ J[u^o(\cdot), x(t'),t'] \qquad (4.3.28)$$

where

$$u(t) = u^*(t) \qquad t_o \leq t \leq t'$$
$$\qquad = u^o(t) \qquad t' < t \leq t_f. \tag{4.3.29}$$

Note that because of (4.3.27) the integral on the right hand side of (4.3.28) is finite for each t' in $(t_o, t_f]$ and that by assumption, (4.3.28) is non-negative. However, if it were true that $x^T(t')S(t')x(t') \to -\infty$ as $t' \to t_e$ for some $x(t')$ then (4.3.28) would tend to minus infinity as $t' \to t_e$. Thus we conclude that $x^T(t')S(t')x(t')$ cannot tend to $-\infty$. Furthermore, it is clear that $J[u^o(\cdot), x(t'), t']$ is bounded above by $J'[\hat{u}(\cdot), x(t'), t']$ where $\hat{u}(\cdot)$ is any given piecewise continuous function defined on $[t', t_f]$. Hence $x^T(t')S(t')x(t')$ cannot tend to $+\infty$ for some $x(t')$ as $t' \to t_e$. Since, by our controllability assumption, $x(t')$ is arbitrary, $||x(t')|| = 1$, we conclude that $S(t')$ cannot tend to infinity as $t \to t_e$. In other words, $S(t)$ exists for all t in $(t_o, t_f]$. We leave it as an exercise for the reader to verify that, in fact, $S(t)$ exists for all t in $[t_o, t_f]$ (see Gelfand and Fomin, 1963).

4.4 Strong Positivity and the Totally Singular Second Variation

In Section 4.3 we demonstrated that if the matrix

Riccati equation has a solution for all t in $[t_o, t_f]$ then $J[u(\cdot)]$ is not only positive definite but also strongly positive. Clearly, in a finite dimensional vector space positive definiteness is equivalent to strong positivity, but in our space of piecewise continuous control functions this is not so. In this section we illustrate the difference between positive definiteness and strong positivity by means of a simple example. In addition, the example illustrates the fact, see Jacobson and Speyer (1971) and Tait (1965) for general proofs, that the totally singular second variation *cannot* be strongly positive. This is of course consistent with the fact that in the totally singular case the matrix Riccati equation is undefined (because "$R^{-1} = \infty$").

Before presenting the example we define "totally singular" precisely as follows:

Definition 4.5 $J[u(\cdot)]$ is said to be totally singular if

$$R(t) = 0 \qquad \text{for all } t \text{ in } [t_o, t_f]. \tag{4.4.1}$$

Now, we consider the following totally singular functional

$$J[u(\cdot)] = \int_o^{t_f} x^2 dt \tag{4.4.2}$$

subject to

$$\dot{x} = u \; ; \qquad x(o) = 0 \tag{4.4.3}$$

$$u(\cdot) \text{ is a member of } U. \tag{4.4.4}$$

Clearly $J[u(\cdot)]$ is positive definite.

Set

$$u(t) = \cos\omega t, \qquad 0 \leq t \leq t_f \tag{4.4.5}$$

so that

$$x(t) = \frac{1}{\omega}\sin \omega t, \qquad 0 \leq t \leq t_f. \tag{4.4.6}$$

With this choice of control $J[u(\cdot)]$ becomes

$$J[u(\cdot)] = \int_o^{t_f} \frac{1}{\omega^2}\sin^2\omega t \; dt \tag{4.4.7}$$

and

$$||u(\cdot)||^2 = \int_o^{t_f} u^2 dt = \int_o^{t_f} \cos^2\omega t dt = \int_o^{t_f} (1-\sin^2\omega t) dt. \tag{4.4.8}$$

By definition, if $J[u(\cdot)]$ were strongly positive then for some $k > 0$ and all $\omega \geq 0$,

$$\frac{1}{\omega^2}\int_o^{t_f} \sin^2\omega t dt \geq k \int_o^{t_f} (1-\sin^2\omega t) dt \tag{4.4.9}$$

$$\text{i.e.} \quad (k + \frac{1}{\omega^2}) \int_o^{t_f} \sin^2\omega t dt \geq k t_f. \tag{4.4.10}$$

But this is impossible because the left hand side of (4.4.10) tends to $\frac{1}{2}kt_f$ as $\omega \to \infty$. In other words, (4.4.2) is not strongly positive.

Note, however, that the functional

$$J[u(\cdot)] = \int_0^{t_f} (x^2 + u^2)dt \tag{4.4.11}$$

is strongly positive. This follows directly because

$$x^2 + u^2 \geq u^2 . \tag{4.4.12}$$

The fact that the totally singular second variation cannot be strongly positive implies that in the totally singular case we should seek (necessary and) sufficient conditions only for non-negativity and for positive definiteness of $J[u(\cdot)]$. We begin this task in the next section.

4.5 A General Sufficiency Theorem for the Second Variation

Here we present and prove a general sufficiency theorem for the partially singular case, which is defined precisely as follows:

Definition 4.6 $J[u(\cdot)]$ is said to be partially singular if

$$R(t) \geq 0 \qquad \text{for all } t \text{ in } [t_o, t_f]. \tag{4.5.1}$$

Following the statement and proof of the general theorem we particularise the result to both the non-singular and the singular special cases. Again, we confine our attention to the situation in which there are no terminal constraints. See Jacobson (1971a) for a more general treatment.

Theorem 4.2 A sufficient condition for non-negativity of $J[u(\cdot)]$ is that there exists for all t in $[t_o, t_f]$ a continuously differentiable, symmetric, matrix function of time $P(\cdot)$ such that

$$\begin{pmatrix} \dot{P}+Q+PA+A^TP & (C+B^TP)^T \\ C+B^TP & R \end{pmatrix} \geq 0 \qquad (4.5.2)$$

for all t in $[t_o, t_f]$ and

$$Q_f - P(t_f) \geq 0. \qquad (4.5.3)$$

Proof: First, we note that for any nxn, symmetric, continuously differentiable, matrix function of time $P(\cdot)$, we have that

$$\int_{t_o}^{t_f} \tfrac{1}{2}\{x^TP(Ax+Bu-\dot{x})\}dt = 0. \qquad (4.5.4)$$

Now, adding this identically zero integral to $J[u(\cdot)]$ yields

$$J[u(\cdot)] = \int_{t_o}^{t_f} \{\tfrac{1}{2}x^TQx+u^TCx+\tfrac{1}{2}u^TRu+\tfrac{1}{2}x^TP(Ax+Bu-\dot{x})\}dt$$

$$+ \tfrac{1}{2}x^T(t_f)Q_fx(t_f). \qquad (4.5.5)$$

In view of our assumptions on $u(\cdot)$ and $P(\cdot)$ we can integrate the term $x^TP\dot{x}$ by parts to obtain

$$J[u(\cdot)] = \int_{t_o}^{t_f} \{\tfrac{1}{2}x^T(\dot{P}+Q+PA+A^TP)x+u^T(C+B^TP)x+\tfrac{1}{2}u^TRu\}dt$$

$$+ \tfrac{1}{2}x^T(t_f)Q_fx(t_f)-\tfrac{1}{2}x^T(t_f)P(t_f)x(t_f).$$

$$(4.5.6)$$

By inspection, then, conditions (4.5.2) and (4.5.3) follow.

4.5.1 The Nonsingular Special Case

We now show that if (4.3.1) holds then existence of a solution $S(\cdot)$ to (4.3.2), (4.3.3) implies that there exists $P(\cdot)$ such that (4.5.2), (4.5.3) hold. It then follows from Theorem 4.1 that the conditions of Theorem 4.2 are both necessary and sufficient for positive-definiteness of $J[u(\cdot)]$ in this nonsingular case.

Theorem 4.3 Suppose that an $S(\cdot)$ exists which satisfies (4.3.2), (4.3.3). Then, there exists a $P(\cdot)$ such that the conditions of Theorem 4.2 are satisfied.

Proof: Consider the quadratic form

$$(\alpha_1^T \ \alpha_2^T) \begin{pmatrix} \dot{S}+Q+SA+A^TS & (C+B^TS)^T \\ C+B^TS & R \end{pmatrix} \begin{pmatrix} \alpha_1 \\ \alpha_2 \end{pmatrix} \tag{4.5.7}$$

where $(\alpha_1^T \ \alpha_2^T)$ is a partitioned, arbitrary, (n+m)-row vector. Since R is positive definite the α_2 that minimizes (4.5.7) for fixed, but arbitrary, α_1 is

$$\alpha_2 = -R^{-1}(C+B^TS)\alpha_1 \tag{4.5.8}$$

and the minimum value of (4.5.7), using this expression, is

$$\alpha_1^T[\dot{S}+Q+SA+A^TS-(C+B^TS)^TR^{-1}(C+B^TS)]\alpha_1 . \tag{4.5.9}$$

But this is zero for all α_1, by (4.3.2). Hence we have from (4.5.7) that

$$\begin{pmatrix} \dot{S}+Q+SA+A^TS & (C+B^TS)^T \\ C+B^TS & R \end{pmatrix} \geq 0 \tag{4.5.10}$$

and

$$Q_f - S(t_f) = 0. \tag{4.5.11}$$

Identifying $S(\cdot)$ with $P(\cdot)$ we see that the conditions of Theorem 4.2 are satisfied.

4.5.2 The Totally Singular Special Case

Here we have that (4.4.1) holds. As a consequence, the conditions of Theorem 4.2 become

$$C+B^TP = 0, \qquad \text{for all } t \text{ in } [t_o,t_f] \quad (4.5.12)$$

$$\dot{P}+Q+PA+A^TP \geq 0, \qquad \text{for all } t \text{ in } [t_o,t_f] \quad (4.5.13)$$

and

$$Q_f - P(t_f) \geq 0 \ . \qquad (4.5.14)$$

As in the nonsingular case, it is possible to obtain a Riccati-type differential equation which, if it has a solution, implies the existence of $P(\cdot)$. In deriving this Riccati equation we assume that B and C are once continuously differentiable with respect to time on $[t_o,t_f]$. Furthermore, we introduce the following, rather important, assumption.

<u>Assumption 4.4</u> The first generalized Legendre-Clebsch necessary condition for optimality is satisfied in strong form for all t in $[t_o,t_f]$. That is,

$$\frac{\partial}{\partial u}\dot{H}_u = 0, \qquad t_o \leq t \leq t_f \qquad (4.5.15)$$

which implies

$$CB \text{ symmetric}, \quad \text{for all } t \text{ in } [t_o, t_f] \tag{4.5.16}$$

and

$$-1\frac{\partial}{\partial u}\ddot{H}_u > 0, \quad \text{for all } t \text{ in } [t_o, t_f] \tag{4.5.17}$$

which implies

$$-B^TQB+\dot{C}B+CAB-\dot{B}^TC^T+B^TA^TC^T < 0 \tag{4.5.18}$$

where

$$H = \tfrac{1}{2}x^TQx+u^TCx+\lambda^T(Ax+Bu) \tag{4.5.19}$$

and

$$-\dot{\lambda} = Qx+A^T\lambda+C^Tu \; . \tag{4.5.20}$$

We shall require the following preliminary lemmas.

Lemma 4.2 The matrices CB and $\dot{C}B+C\dot{B}$ are symmetric.

Proof: From (4.5.15) and (4.5.19), (4.5.20) we have that

$$\frac{\partial}{\partial u}\dot{H}_u = \frac{\partial}{\partial u}\left[\frac{d}{dt}(B^T\lambda+Cx)\right] \tag{4.5.21}$$

$$= \frac{\partial}{\partial u}[\dot{B}^T\lambda+B^T\dot{\lambda}+\dot{C}x+CAx+CBu] \tag{4.5.22}$$

$$= \frac{\partial}{\partial u}[\dot{B}^T\lambda-B^TQx-B^TA^T\lambda-B^TC^Tu+\dot{C}x+CAx+CBu] \tag{4.5.23}$$

$$= CB - B^TC^T \tag{4.5.24}$$

$$= 0 \quad \text{for all } t \text{ in } [t_o,t_f]. \tag{4.5.25}$$

That is, CB is symmetric. Now, as C and B are continuously differentiable this implies that the lemma is proved.

<u>Lemma 4.3</u> $\frac{\partial}{\partial u}\ddot{H}_u = -B^TQB+\dot{C}B+CAB-\dot{B}^TC^T+B^TA^TC^T$. (4.5.26)

Proof: By direct differentiation and Lemma 4.2.

<u>Theorem 4.4</u> Suppose that there exists a function S(·) which satisfies for all t in $[t_o,t_f]$, the equations

$$-\dot{S}=Q+SA+A^TS+[(AB-\dot{B})^TS+B^TQ-CA-\dot{C}]^T[\frac{\partial}{\partial u}\ddot{H}_u]^{-1}$$

$$[(AB-\dot{B})^TS+B^TQ-CA-\dot{C}] \tag{4.5.27}$$

$$C(t_f)+B^T(t_f)S(t_f) = 0 \tag{4.5.28}$$

$$Q_f - S(t_f) \geq 0 . \tag{4.5.29}$$

Then, there exists a P(·) which satisfies (4.5.12) - (4.5.14).

Proof: From (4.5.17) and (4.5.27) we see that

$$\dot{S}+Q+SA+A^TS \geq 0, \quad \text{for all } t \text{ in } [t_o,t_f]. \tag{4.5.30}$$

Next, premultiplying (4.5.27) by B^T and using Lemmas 4.2, 4.3, we obtain

$$-B^T\dot{S} = (C+B^TS)A+\dot{C}+\dot{B}^TS$$

$$+ [B^TS(AB-\dot{B})+C(AB-\dot{B})][\frac{\partial}{\partial u}\ddot{H}_u]^{-1}$$

$$[(AB-\dot{B})^TS+B^TQ-CA-\dot{C}] \qquad (4.5.31)$$

which, re-arranged, is

$$-\frac{d}{dt}(C+B^TS)=(C+B^TS).\{A+(AB-\dot{B})[\frac{\partial}{\partial u}\ddot{H}_u]^{-1}$$

$$[(AB-\dot{B})^TS+B^TQ-CA-\dot{C}]\}. \qquad (4.5.32)$$

Now this is an ordinary linear homogeneous differential equation for $(C+B^TS)$ which has boundary condition (4.5.28). Consequently,

$$C+B^TS = 0, \quad \text{for all } t \text{ in } [t_o,t_f]. \qquad (4.5.33)$$

Identifying $P(\cdot)$ with $S(\cdot)$ we see that the theorem is true.

4.6 Necessary and Sufficient Conditions for Non-negativity of the Totally Singular Second Variation

We have seen via Theorem 4.3 that, in the nonsingular case, Theorem 4.2 provides conditions which are both necessary and sufficient for positive-definiteness of $J[u(\cdot)]$. As a consequence of the necessity of conditions (4.5.27-29) (McDanell and Powers, 1970) subject, of course, to Assumption 4.4, we have necessity and sufficiency also in the totally singular case (see also Speyer and Jacobson (1971)). The question now arises as to the possible necessity and sufficiency of conditions (4.5.2) and (4.5.3) in the totally singular case, in the absence of Assumption 4.4 and the continuous differentiability of B and C. It turns out that a slightly more abstract version of these conditions is both necessary and sufficient (Jacobson and Speyer, 1971) and it is this set of conditions which occupies us in this section.

Theorem 4.5 i) Necessary Condition. Under assumptions 4.1-4.3, a necessary condition for $J[u(\cdot)]$ to be non-negative in the totally singular case is that there exists for all t in $(t_o,t_f]$ a real symmetric nxn matrix function of time $\hat{P}(\cdot)$ which is monotone increasing† in t such that

$$C+B^T P = 0, \quad \text{for all } t \text{ in } (t_o,t_f] \tag{4.6.1}$$

$$Q_f-P(t_f) = -\hat{P}(t_f) \geq 0 \tag{4.6.2}$$

where

$$P = S+\phi^T\hat{P}\phi, \quad \text{for all } t \text{ in } (t_o,t_f] \tag{4.6.3}$$

† i.e., $z^T\hat{P}(t)z$ is monotone increasing in t, where z belongs to R^n and is fixed but arbitrary.

$$\dot{\phi} = -\phi A, \qquad \phi(t_f) = I \tag{4.6.4}$$

and

$$-\dot{S} = Q+SA+A^T S, \qquad S(t_f) = Q_f \tag{4.6.5}$$

ii) Sufficient Condition. In addition to the above stated conditions, $\hat{P}(\cdot)$ exists for all t in $[t_o,t_f]$ (strengthened existence condition).

Note that the gap between the necessary and the sufficient condition is minimal.

Before proceeding to the proof of this main theorem, we verify that (4.5.12)-(4.5.14) are implied by (4.6.1)-(4.6.5) in the case that $\hat{P}(\cdot)$ is continuously differentiable. Clearly, (4.6.1) is just condition (4.5.12) and (4.6.2) is condition (4.5.14). Now, differentiating (4.6.3) with respect to time yields

$$\dot{P} = \dot{S} + \dot{\phi}^T\hat{P}\phi + \phi^T\dot{\hat{P}}\phi + \phi^T\hat{P}\dot{\phi} \tag{4.6.6}$$

which is

$$\dot{P}+Q+SA+A^TS+A^T\phi^T\hat{P}\phi-\phi^T\dot{\hat{P}}\phi+\phi^T\hat{P}\phi A = 0 \tag{4.6.7}$$

i.e.,

$$\dot{P}+Q+(S+\phi^T\hat{P}\phi)A+A^T(S+\phi^T\hat{P}\phi) - \phi^T\dot{\hat{P}}\phi = 0 \tag{4.6.8}$$

which, because of the monotonicity of $\hat{P}$ and the

expression (4.6.3) implies

$$\dot{P}+Q+PA+A^TP \geq 0 \tag{4.6.9}$$

which is (4.5.13).

The question now arises as to whether or not there are problems for which the conditions of Theorem 4.5 are satisfied, but for which no continuously differentiable $\hat{P}(\cdot)$ exists. We supply such an example below:

$$J[u(\cdot)] = \int_{t_o}^{t_f} uCx\,dt \tag{4.6.10}$$

subject to

$$\dot{x} = u \quad ; \quad x(t_o) = 0 \tag{4.6.11}$$

where we assume that $C(\cdot)$ is monotone decreasing in t and that

$$C(t_f) \geq 0 \; . \tag{4.6.12}$$

It is clear that

$$P = -C, \quad \text{for all } t \text{ in } [t_o,t_f] \tag{4.6.13}$$

$$P = \hat{P} \tag{4.6.14}$$

satisfy the conditions of Theorem 4.5. However, if C is not continuously differentiable there does not exist a continuously differentiable function $P(\cdot)$

which satisfies (4.5.12)-(4.5.14). In other words, for this example, the more abstract conditions of Theorem 4.5 apply whereas the general sufficiency Theorem 4.2 is inapplicable.

We commence the proof of Theorem 4.5 with a sequence of Lemmas.

Lemma 4.4 The totally singular quadratic functional is expressible in the equivalent canonical form

$$J[u(\cdot)] = \int_{t_o}^{t_f} u^T D y\,dt \tag{4.6.15}$$

subject to

$$\dot{y} = Eu \; ; \qquad y(t_o) = 0 \tag{4.6.16}$$

where

$$D = [C+B^T S]\phi^{-1} \tag{4.6.17}$$

$$E = \phi B \tag{4.6.18}$$

$$\dot{\phi} = -\phi A; \qquad \phi(t_f) = I \tag{4.6.19}$$

$$-\dot{S} = Q+SA+A^T S; \qquad S(t_f) = Q_f \tag{4.6.20}$$

and where

$$y(t) = \phi(t)x(t), \quad \text{for all } t \text{ in } [t_o,t_f]. \tag{4.6.21}$$

Proof: Adjoin (4.2.2) to (4.2.1) using a continuously differentiable vector multiplier $\lambda(\cdot)$ to form

$$J[u(\cdot)] = \int_{t_o}^{t_f} [\tfrac{1}{2}x^TQx+u^TCx+\lambda^T(Ax+Bu-\dot{x})]dt + \tfrac{1}{2}x^T(t_f)Q_fx(t_f). \quad (4.6.22)$$

Integrating by parts and setting

$$\lambda(t) = \tfrac{1}{2}S(t)x(t) \quad (4.6.23)$$

yields

$$J[u(\cdot)] = \int_{t_o}^{t_f} [\tfrac{1}{2}x^T(\dot{S}+Q+SA+A^TS)x + u^T(C+B^TS)x]dt + \tfrac{1}{2}x^T(t_f)[Q_f-S(t_f)]x(t_f). \quad (4.6.24)$$

Now, using (4.6.20) in (4.6.24) yields

$$J[u(\cdot)] = \int_{t_o}^{t_f} u^T(C+B^TS)xdt. \quad (4.6.25)$$

Differentiating (4.6.21) with respect to time we obtain easily the result that

$$J[u(\cdot)] = \int_{t_o}^{t_f} u^T(C+B^TS)\phi^{-1}ydt \quad (4.6.26)$$

subject to

$$\dot{y} = \phi B u \; ; \quad y(t_o) = 0 \; . \tag{4.6.27}$$

For the converse, we note that because ϕ is invertible (4.6.26) is just

$$J[u(\cdot)] = \int_{t_o}^{t_f} u^T(C+B^T S)x dt \tag{4.6.28}$$

and

$$\dot{x} = Ax + Bu \; ; \quad x(t_o) = 0 \; . \tag{4.6.29}$$

Now, adjoining (4.6.29) to (4.6.28) with a multiplier function $\tilde{\lambda}(\cdot)$, integrating by parts and setting

$$\tilde{\lambda}(t) = \tfrac{1}{2}\tilde{S}(t)x(t) \tag{4.6.30}$$

yields

$$J[u(\cdot)] = \int_{t_o}^{t_f} [\tfrac{1}{2}x^T(\dot{\tilde{S}}+\tilde{S}A+A^T\tilde{S})x+u^T(C+B^T S+B^T\tilde{S})x]dt$$

$$- \tfrac{1}{2}x^T(t_f)\tilde{S}(t_f)x(t_f) \; . \tag{4.6.31}$$

Setting

$$\tilde{S}(\cdot) = -S(\cdot) \tag{4.6.32}$$

yields

$$J[u(\cdot)] = \int_{t_o}^{t_f} (\tfrac{1}{2}x^TQx+u^TCx)dt+\tfrac{1}{2}x^T(t_f)Q_fx(t_f) \quad (4.6.33)$$

so that the lemma is proved.

Lemma 4.5 Condition (i) of Theorem 4.5 is equivalent to the existence for all t in $(t_o,t_f]$ of a real symmetric monotone increasing matrix function of time $\hat{P}(\cdot)$ such that

$$D+E^T\hat{P} = 0, \quad \text{for all t in } (t_o,t_f] \quad (4.6.34)$$

$$-\hat{P}(t_f) \geq 0 . \quad (4.6.35)$$

Proof: Let

$$P = \phi^T\hat{P}\phi + S \quad (4.6.36)$$

and substitute into the conditions of Theorem 4.5 and the conditions (4.6.34), (4.6.35).

We now define the related nonsingular functional

$$J_N[u(\cdot),\varepsilon] = J[u(\cdot)] + \frac{1}{2\varepsilon}\int_{t_o}^{t_f} u^Tudt, \quad \varepsilon>0 \quad (4.6.37)$$

i.e.

$$J_N[u(\cdot),\varepsilon] = \int_{t_o}^{t_f}[u^TDy + \frac{1}{2\varepsilon}u^Tu]dt \quad (4.6.38)$$

subject to

$$\dot{y} = Eu \; ; \qquad y(t_o) = 0 \qquad (4.6.39)$$

which leads to the following lemmas.

Lemma 4.6 If $J[u(\cdot)]$ is non-negative then $J_N[u(\cdot),\varepsilon]$ is positive definite. The proof is obvious.

Lemma 4.7 If $J[u(\cdot)]$ is non-negative then the matrix Riccati equation

$$-\dot{S}_\varepsilon = -(D+E^TS_\varepsilon)^T(D+E^TS_\varepsilon)\varepsilon \quad ; \quad S_\varepsilon(t_f) = 0 \quad (4.6.40)$$

associated with $J_N[u(\cdot),\varepsilon]$ has a solution which exists for all t in $[t_o,t_f]$.

Proof: See Theorem 4.1 and Gelfand and Fomin (1963).

Lemma 4.8 If the matrix Riccati equation (4.6.40) has a solution which exists in the interval $[\tau,t_f]$, $t_o \leq \tau \leq t_f$ then the control function that minimizes

$$J_N^{\tau}[y(\tau),u(\cdot),\varepsilon,\tau] = \int_\tau^{t_f}[u^TDy + \frac{1}{2\varepsilon}u^Tu]dt \quad (4.6.41)$$

subject to

$$\dot{y} = Eu \; ; \quad y(\tau) \text{ given} \qquad (4.6.42)$$

is

$$u^o(t) = -\varepsilon[D(t)+E^T(t)S_\varepsilon(t)]y(t), \qquad t \text{ in } [\tau,t_f] \qquad (4.6.43)$$

and moreover,

$$\min_{u(\cdot)} J_N{}^{\tau}[y(\tau),u(\cdot),\varepsilon,\tau] = J^{o}[y(\tau),\varepsilon,\tau] = \tfrac{1}{2}y^{T}(\tau)S_{\varepsilon}(\tau)y(\tau) \tag{4.6.44}$$

Proof: See Theorem 4.1.

Lemma 4.9 $S_{\varepsilon}(\cdot)$ is a continuous function of the parameter ε.

Proof: If $J[u(\cdot)]$ is non-negative it follows that $S_{\varepsilon}(\cdot)$ exists for all ε. Furthermore, the right hand side of (4.6.40) is analytic in S_{ε} and ε(Coddington and Levinson, 1955).

Lemma 4.10 $J^{o}[y(\tau),\varepsilon,\tau]$ is a monotone decreasing function of ε. Hence $S_{\varepsilon}(\tau)$ is, by definition, a monotone decreasing matrix function of ε.

Proof: For some arbitrary $y(\tau)$, ε_1, τ we have

$$J^{o}[y(\tau),\ \varepsilon_1,\ \tau] = \min_{u(\cdot)} \int_{\tau}^{t_f} [u^{T}Dy + \frac{1}{2\varepsilon_1}\ u^{T}u]dt. \tag{4.6.45}$$

Let the control which minimizes (4.6.45) be denoted $u_1(\cdot)$ and its associated trajectory be $y_1(\cdot)$. Then, for any $\varepsilon_2 \geq \varepsilon_1$ it is clear that

$$\int_{\tau}^{t_f} [u_1{}^{T}Dy_1 + \frac{1}{2\varepsilon_2}\ u_1{}^{T}u_1]dt \leq \int_{\tau}^{t_f} [u_1{}^{T}Dy_1 + \frac{1}{2\varepsilon_1}\ u_1{}^{T}u_1]dt \tag{4.6.46}$$

and by definition,

$$J^o[y(\tau),\varepsilon_2,\tau] \leq \int_\tau^{t_f} [u_1^T D y_1 + \frac{1}{2\varepsilon_2} u_1^T u_1] dt. \quad (4.6.47)$$

Thus, for any $\varepsilon_2 \geq \varepsilon_1$, we have

$$J^o[y(\tau),\varepsilon_2,\tau] \leq J^o[y(\tau),\varepsilon_1,\tau]. \quad (4.6.48)$$

As $y(\tau)$ and τ are arbitrary, the lemma is proved.

Lemma 4.11 Under Assumptions 4.1 and 4.3, if $J[u(\cdot)]$ is non-negative then $S_\infty(t) = \lim_{\varepsilon\to\infty} S_\varepsilon(t)$ exists for all t in $(t_o,t_f]$ and is negative semi-definite.

Proof: From (4.6.40), $S_\varepsilon(\tau) \leq 0$ for all τ in $[t_o,t_f]$, for all ε, $0 < \varepsilon < \infty$.

By Lemma 4.10, $S_\varepsilon(\tau)$ is a monotone decreasing function of ε so that it has a limit (possibly $-\infty$). Now, given an arbitrary time τ in the interval (t_o,t_f) we can, by Assumption 4.3, construct $u_3(t)$, $t_o \leq t \leq \tau$ such that

$$\int_{t_o}^{\tau} [u_3^T D y_3 + \frac{1}{2\varepsilon} u_3^T u_3] dt < \infty \quad (4.6.49)$$

and such that

$$y_3(\tau) = \gamma(\tau), \qquad \gamma(\tau) \text{ arbitrary}. \quad (4.6.50)$$

Suppose that $||S_\varepsilon(\tau)|| \to \infty$ as $\varepsilon \to \infty$. Then, by Lemmas 4.9, 4.10, 4.11, $J^o[y_3(\tau),\varepsilon,\tau]$ can be made large and negative for some $\gamma(\tau)$ and for some ε sufficiently large. This implies that one can make the following inequality hold, for ε sufficiently large,

$$\int_{t_o}^{\tau} \{u_3^T D y_3 + \frac{1}{2\varepsilon} u_3^T u_3\} dt + J^o[y_3(\tau), \varepsilon, \tau] < 0. \quad (4.6.51)$$

But this contradicts the fact that $J[u(\cdot)]$ is non-negative, so $||S_\varepsilon(\tau)|| \not\to \infty$ as $\varepsilon \to \infty$. Since τ is arbitrary and since $S_\varepsilon(t_f) = 0$ for all ε we conclude that $S_\infty(\tau)$ exists for all τ in $(t_o, t_f]$.

Lemma 4.12 If $S_\infty(\tau)$ exists it is a monotone increasing function of τ.

Proof: From (4.6.40),

$$S_\infty(\tau) = \lim_{\varepsilon\to\infty}\left[\int_{t_f}^{\tau} (D+E^T S_\varepsilon)^T (D+E^T S_\varepsilon)\varepsilon dt\right] \quad (4.6.52)$$

and, because of the existence of each limit, the right hand side of (4.6.52) becomes

$$\lim_{\varepsilon\to\infty}\left[\int_{t_f}^{\tau+\Delta} (D+E^T S_\varepsilon)^T (D+E^T S_\varepsilon)\varepsilon dt\right]$$

$$+ \lim_{\varepsilon\to\infty}\left[\int_{\tau+\Delta}^{\tau} (D+E^T S_\varepsilon)^T (D+E^T S_\varepsilon)\varepsilon dt\right]. \quad (4.6.53)$$

That is,

$$S_\infty(\tau) = S_\infty(\tau+\Delta) + \lim_{\varepsilon\to\infty}\int_{\tau+\Delta}^{\tau}[(D+E^T S_\varepsilon)^T (D+E^T S_\varepsilon)\varepsilon dt] \quad (4.6.54)$$

and the lemma follows.

Lemma 4.13 If $S_\infty(\tau)$ exists for all τ in $(t_o,t_f]$ then

$$D(\tau)+E^T(\tau)S_\infty(\tau) = 0 \quad \text{a.e. in } [t_o,t_f]. \tag{4.6.55}$$

Proof: Suppose the contrary. Then, for some τ in (t_o,t_f), by Lemma 4.9, we have that there exists $\varepsilon^* > 0$ such that

$$\left|\left|\int_{t_f}^{\tau} (D+E^TS_\varepsilon)^T(D+E^TS_\varepsilon)dt\right|\right| \geq \rho > 0, \quad \text{for all } \varepsilon \geq \varepsilon^* \tag{4.6.56}$$

so that

$$\lim_{\varepsilon\to\infty}\left|\left|\int_{t_f}^{\tau} (D+E^TS_\infty)^T(D+E^TS_\infty)\varepsilon dt\right|\right| = \infty \tag{4.6.57}$$

which contradicts the fact that $S_\infty(\tau)$ exists, $t_o < \tau \leq t_f$.

We are now in a position to prove Theorem 4.5.

Proof of Theorem 4.5 i) Necessary Condition. If $J[u(\cdot)]$ is non-negative then by Lemmas 4.11-4.13 there exists $S_\infty(\tau)$, $t_o < \tau \leq t_f$ which is a real symmetric monotone increasing matrix function of time such that

$$D + E^TS_\infty = 0 \qquad \text{a.e. in } [t_o,t_f] \tag{4.6.58}$$

and

$$S_\infty(t_f) = 0 . \tag{4.6.59}$$

Now, since $S_\infty(\cdot)$ is a monotone function, we can define

$$\hat{P}(t_f) = \lim_{t\to t_f} S_\infty(t) \tag{4.6.60}$$

and consequently we have

$$-\hat{P}(t_f) \geq 0 . \tag{4.6.61}$$

Furthermore, defining

$$\hat{P}(t) = S_\infty(t), \quad t_o < t < t_f \tag{4.6.62}$$

yields

$$D+E^T\hat{P} = 0 \qquad \text{a.e. in } [t_o,t_f]. \tag{4.6.63}$$

where $\hat{P}(t)$ is defined for all t in $(t_o,t_f]$.

In order to establish the necessary conditions we need only show that

$$D+E^T\hat{P} = 0, \qquad \text{for all t in } (t_o,t_f]. \tag{4.6.64}$$

In order to do this, suppose that for some t in (t_o,t_f)

$$D+E^T\hat{P} \neq 0 . \tag{4.6.65}$$

As D, E are continuous in t, (4.6.65) can occur only if a jump in $\hat{P}$ occurs at time t that does not lie in the null space of $E^T(t)$. Moreover, as $\hat{P}$ is monotone increasing in t, it follows that

$$DE + E^T\hat{P}E > 0 . \tag{4.6.66}$$

Again, since $\hat{P}$ is monotone and D and E are continuous we have that (4.6.66) holds during a time interval $[t, t+\Delta]$, $\Delta>0$, contradicting (4.6.63). Hence,

$$D + E^T\hat{P} = 0, \quad \text{for all } t \text{ in } (t_o,t_f) . \tag{4.6.67}$$

Equation (4.6.64) now follows because of (4.6.60).

ii) Sufficient Condition. Suppose now that $\hat{P}(t)$ exists for all t in $[t_o,t_f]$ (strengthened existence condition). Adjoin the dynamics (4.6.16) to $J[u(\cdot)]$ as follows.

$$J[u(\cdot)] = \int_{t_o}^{t_f}[u^TDy + y^T\hat{P}(Eu-\dot{y})]dt \tag{4.6.68}$$

$$= \int_{t_o}^{t_f} u^T(D+E^T\hat{P})y\ dt - \int_{t_o}^{t_f} y^T\hat{P}\dot{y}\ dt . \tag{4.6.69}$$

The first integral is zero because of (4.6.34). The remaining integral can be written in Stieltjes form as

$$-\int_{t_o}^{t_f} y^T\hat{P}dy \tag{4.6.70}$$

which, upon integration by parts, becomes

$$\int_{t_o}^{t_f} \tfrac{1}{2}y^T d\hat{P}y - \tfrac{1}{2}y^T\hat{P}y\Big|_{t_o}^{t_f} \qquad (4.6.71)$$

which is non-negative owing to the monotonicity of $\hat{P}(\cdot)$ and the fact that $-\hat{P}(t_f) \geq 0$ and $y(t_o) = 0$. This concludes the proof of Theorem 4.5.

It is worth noting that an extension of Theorem 4.5 is known, for the case where terminal equality constraints are present in the formulation of the optimal control problem (and hence in the second variation), (Jacobson and Speyer, 1971).

4.7 Necessary Conditions for Optimality

As Theorem 4.5 provides necessary and sufficient conditions for non-negativity of the second variation, one would expect to be able to derive known necessary conditions, for the totally singular case, from these. This is indeed possible, as the next several results show.

Theorem 4.6 (Robbins, 1967; Goh, 1966) A necessary condition for $J[u(\cdot)]$ to be non-negative is that CB be symmetric for all t in $[t_o, t_f]$.

Proof: From (4.6.1)

$$CB + B^T PB = 0, \qquad \text{for all } t \text{ in } (t_o, t_f]. \qquad (4.7.1)$$

Since P is symmetric it follows that CB is symmetric

for all t in $[t_o,t_f]$. The theorem follows from the continuity of CB with respect to time.

Theorem 4.7 (Jacobson, 1969; Gabasov, 1968) A necessary condition for $J[u(\cdot)]$ to be non-negative is that

$$CB + B^TSB \geq 0, \quad \text{for all t in } [t_o,t_f] \quad (4.7.2)$$

where

$$-\dot{S} = Q+SA+A^TS \ ; \quad S(t_f) = Q_f \ . \quad (4.7.3)$$

Proof: From (4.6.1) and (4.6.3),

$$(C+B^TP)B = (C+B^TS+B^T\phi^T\hat{P}\phi)B = 0, \quad (4.7.4)$$

$$\text{for all t in } (t_o,t_f].$$

As $\hat{P}(t) \leq 0$, for all t in $[t_o,t_f]$ it follows from (4.7.4) that

$$CB + B^TSB \geq 0, \quad \text{for all t in } (t_o,t_f]. \quad (4.7.5)$$

By continuity of C, B and S, inequality (4.7.2) follows.

Theorem 4.8 (Kelley et al., 1967; Robbins, 1967; Tait, 1965, generalized Legendre-Clebsch condition) A necessary condition for $J[u(\cdot)]$ to be non-negative is that

$$(-1)\frac{\partial}{\partial u}\ddot{H}_u \geq 0 \ , \quad \text{for all t in } [t_o,t_f] \quad (4.7.6)$$

where H is defined by (4.5.19).

Proof: From (4.6.1),

$$\int_{t_1}^{t_2} d[(C+B^T P)B] = 0 \tag{4.7.7}$$

for all t_1, t_2 in $(t_o, t_f]$, $t_1 < t_2$.

which is

$$\int_{t_1}^{t_2} (\dot{C}B+C\dot{B})\,dt + \int_{t_1}^{t_2} (\dot{B}^T PB+B^T P\dot{B})\,dt + \int_{t_1}^{t_2} B^T dPB = 0. \tag{4.7.8}$$

Using (4.6.3) in (4.7.8) yields

$$\int_{t_1}^{t_2} (\dot{C}B+C\dot{B}+\dot{B}^T PB+B^T P\dot{B})\,dt - \int_{t_1}^{t_2} (B^T QB+B^T PAB+B^T A^T PB)\,dt$$

$$+ \int_{t_1}^{t_2} B^T \phi^T d\hat{P}\phi B = 0. \tag{4.7.9}$$

Using (4.6.1) in (4.7.9) and collecting terms yields,

$$\int_{t_1}^{t_2} \frac{\partial}{\partial u}\ddot{H}_u\,dt + \int_{t_1}^{t_2} B^T \phi^T d\hat{P}\phi B = 0. \tag{4.7.10}$$

Since $\hat{P}$ is monotone increasing in t we obtain the result that

$$(-1)\frac{\partial}{\partial u}\ddot{H}_u \geq 0, \quad \text{for all } t \text{ in } (t_o, t_f). \tag{4.7.11}$$

The continuity with respect to time of the left hand side of (4.7.11) yields (4.7.6).

4.8 Other Necessary and Sufficient Conditions

The necessary and sufficient conditions presented and proved in Section 4.6 do not require that (4.5.17) hold. However, if this strong form of the generalized Legendre-Clebsch condition is assumed then either Goh's or Kelley's transformation technique can be used to transform the totally singular second variation into a nonsingular one. In the case of Goh's transformation the dimension of the state space is preserved and the ensuing analysis (Goh, 1970; McDanell and Powers, 1970) results in the Riccati differential equation of (4.5.27). Application of Kelley's transformation technique yields a nonsingular problem in a reduced dimensional state space and hence a Riccati differential equation of reduced dimension results (Speyer and Jacobson, 1971). Both these Riccati equations imply that the second variation is strongly positive with respect to the control variable in the transformed spaces.

Note that our conditions, Theorem 4.5, have been extended recently to the partially singular case by Anderson (1973).

4.9 Sufficient Conditions for a Weak Local Minimum

As pointed out in the introduction to this chapter, a sufficient condition for a weak local minimum in the nonsingular optimal control problem is that a solution exists for all t in $[t_o, t_f]$ to the matrix Riccati equation (4.3.2). In other words, a sufficient condition is that $J[u(\cdot)]$ given by (4.2.1) be strongly positive, where

$$\begin{aligned} Q(t) &= H_{xx}(t) \\ C(t) &= H_{ux}(t) \\ R(t) &= H_{uu}(t) \\ Q_f &= F_{xx}(t_f) \end{aligned} \tag{4.9.1}$$

and where

$$\begin{aligned} A(t) &= f_x(t) \\ B(t) &= f_u(t) \ . \end{aligned} \tag{4.9.2}$$

In the singular case the second variation cannot be strongly positive, but sufficiency is nevertheless ensured by the strong positivity of the transformed second variation (Goh, 1970; Jacobson and Speyer, 1971; McDanell and Powers, 1970; Speyer and Jacobson, 1971).

A recent advance in the theory of singular optimal control problems is provided by Moyer (1973) who gives sufficient, but generally not necessary, conditions for a strong minimum. This is accomplished by a field theory.

4.10 Existence Conditions for the Matrix Riccati Differential Equation

This chapter is concerned mainly with (necessary and) sufficient conditions for non-negativity of the singular second variation. It is an interesting fact that the sufficient conditions (4.5.12)-(4.5.14) play an important role in deducing conditions which guarantee existence of a solution to (4.3.2-3). The conditions so deduced (Jacobson, 1970b) are less restrictive than those known heretofore (Kalman, 1960; Breakwell and Ho, 1965) which we state here without proof.

Theorem 4.9 Sufficient conditions for the existence of $S(t)$, $t_o \leq t \leq t_f$ which satisfies

$$-\dot{S} = Q+SA+A^TS-(C+B^TS)^TR^{-1}(C+B^TS) \tag{4.10.1}$$

$$S(t_f) = Q_f \tag{4.10.2}$$

are that

$$Q - C^TR^{-1}C \geq 0, \quad \text{for all } t \text{ in } [t_o,t_f] \tag{4.10.3}$$

$$R^{-1} > 0, \quad \text{for all } t \text{ in } [t_o,t_f] \tag{4.10.4}$$

$$Q_f \geq 0 \,. \tag{4.10.5}$$

Clearly, (4.10.1) can be written in the form

$$-\dot{S} = \tilde{Q}+S\tilde{A}+\tilde{A}^TS-SBR^{-1}B^TS \tag{4.10.6}$$

where

$$\tilde{Q} = Q-C^TR^{-1}C \tag{4.10.7}$$

$$\tilde{A} = A - BR^{-1}C \ . \tag{4.10.8}$$

In view of this, we shall study the question of existence of a solution to (4.10.6) but, for simplicity of notation we shall simply refer to $\tilde{A}$ and $\tilde{Q}$ as A and Q.

We then have the following theorem.

Theorem 4.10 A sufficient condition for the existence of $S(t)$, $t_o \le t \le t_f$ which satisfies

$$-\dot{S} = Q+SA+A^TS-SBR^{-1}B^TS \tag{4.10.9}$$

$$S(t_f) = Q_f \tag{4.10.10}$$

is that there exists an nxn symmetric matrix function of time $P(\cdot)$ whose elements are continuously differentiable functions of time in the interval $[t_o,t_f]$ such that

$$B^TP = 0, \quad \text{for all } t \text{ in } [t_o,t_f] \tag{4.10.11}$$

$$\dot{P}+Q+PA+A^TP=M(t) \ge 0, \quad \text{for all } t \text{ in } [t_o,t_f] \tag{4.10.12}$$

$$Q_f - P(t_f) = G_f \geq 0. \tag{4.10.13}$$

Proof: Let

$$\overline{P}(t)+\overline{S}(t)=P(t), \quad \text{for all } t \text{ in } [t_o,t_f] \tag{4.10.14}$$

where $\overline{P}(\cdot)$ and $\overline{S}(\cdot)$ are real symmetric matrix functions of time. Then, from (4.10.11), (4.10.12) and (4.10.14)

$$-\dot{\overline{P}}-\dot{\overline{S}} = Q+A^T(\overline{P}+\overline{S})+(\overline{P}+\overline{S})A-M-(\overline{P}+\overline{S})BR^{-1}B^T(\overline{P}+\overline{S}) \tag{4.10.15}$$

$$= Q+A^T(\overline{P}+\overline{S})+(\overline{P}+\overline{S})A-M-\overline{S}BR^{-1}B^T\overline{S}-\overline{S}BR^{-1}B^T\overline{P}$$

$$-\overline{P}BR^{-1}B^T\overline{S}-\overline{P}BR^{-1}B^T\overline{P}. \tag{4.10.16}$$

Using (4.10.11) and (4.10.14) in (4.10.16) yields

$$-\dot{\overline{P}}-\dot{\overline{S}} = Q+A^T(\overline{P}+\overline{S})+(\overline{P}+\overline{S})A-M-\overline{S}BR^{-1}B^T\overline{S}+\overline{P}BR^{-1}B^T\overline{P}. \tag{4.10.17}$$

Now choose,

$$-\dot{\overline{P}} = -M+A^T\overline{P}+\overline{P}A+\overline{P}BR^{-1}B^T\overline{P} \tag{4.10.18}$$

and

$$\overline{P}(t_f) = -G_f \ . \tag{4.10.19}$$

From (4.10.18), (4.10.19) we see that the variable $(-\overline{P})$ satisfies a Riccati equation for which the conditions of Theorem 4.9 hold, namely

$$M(t) \geq 0, \quad \text{for all } t \text{ in } [t_o, t_f] \tag{4.10.20}$$

$$R^{-1}(t) > 0, \text{ for all } t \text{ in } [t_o, t_f] \tag{4.10.21}$$

$$G_f \geq 0 \tag{4.10.22}$$

so that $\overline{P}(t)$ exists for all t in $[t_o, t_f]$.

Now, using (4.10.18) in (4.10.17) and (4.10.19) in (4.10.13) yields

$$-\dot{\overline{S}} = Q+\overline{S}A+A^T\overline{S}-\overline{S}BR^{-1}B^T\overline{S} \tag{4.10.23}$$

$$\overline{S}(t_f) = Q_f \tag{4.10.24}$$

which are identical to (4.10.9), (4.10.10). Furthermore, because P(t) and $\overline{P}(t)$ exist for all t in $[t_o, t_f]$ it follows from (4.10.14) that $\overline{S}(t)$ exists for all t in $[t_o, t_f]$, and the theorem is proved.

The next theorem and the following example emphasize the validity of Theorem 4.10.

Theorem 4.11 The conditions of Theorem 4.10 are not more stringent than those of Theorem 4.9.

Proof: Writing $Q-C^TR^{-1}C$ for Q and $A-BR^{-1}C$ for A in the conditions of Theorem 4.10 yields

$$B^TP = 0, \quad \text{for all } t \text{ in } [t_o, t_f] \tag{4.10.25}$$

$$\dot{P}+Q-C^TR^{-1}C+(A-BR^{-1}C)^TP+P(A-BR^{-1}C)=M(t) \geq 0 \tag{4.10.26}$$

for all t in $[t_o, t_f]$ and,

$$Q_f - P(t_f) = G_f \geq 0 \, . \tag{4.10.27}$$

Clearly (4.10.25)-(4.10.27) are satisfied with $P(\cdot) \equiv 0$, if the conditions of Theorem 4.9 are satisfied.

The following example illustrates that the conditions of Theorem 4.10 are *less restrictive* than those of Theorem 4.9.

4.10.1 An Example

Let

$$n=2, \; m=1, \; t_o=0, \; t_f=1 \tag{4.10.28}$$

$$C=0, \; B^T=(0,1), \; R^{-1}=1, \; Q_f=0 \tag{4.10.29}$$

$$Q = \begin{pmatrix} -1 & 0 \\ 0 & 4 \end{pmatrix}, \; A = \begin{pmatrix} 0 & 1 \\ 0 & 0 \end{pmatrix}. \tag{4.10.30}$$

Clearly these values do not satisfy the conditions of Theorem 4.9; that is, the known sufficiency conditions are violated. The conditions of Theorem 4.10 become

$$(P_{12} \quad P_{22}) = 0, \quad \text{for all t in } [0,1] \tag{4.10.31}$$

$$\begin{pmatrix} \dot{P}_{11} & 0 \\ 0 & 0 \end{pmatrix} + \begin{pmatrix} 0 & P_{11} \\ P_{11} & 0 \end{pmatrix} + \begin{pmatrix} -1 & 0 \\ 0 & 4 \end{pmatrix} \geq 0, \tag{4.10.32}$$

for all t in [0,1]

and

$$\begin{pmatrix} -P_{11}(1) & 0 \\ 0 & 0 \end{pmatrix} \geq 0 \,. \tag{4.10.33}$$

Let us now choose

$$P_{11}(1) = 0 \tag{4.10.34}$$

which satisfies (4.10.33) with equality.

If we then choose

$$\dot{P}_{11} = 2 \tag{4.10.35}$$

we get

$$P_{11}(0) = -2 \tag{4.10.36}$$

and (4.10.32) becomes

$$\begin{pmatrix} 1 & -2+2t \\ -2+2t & 4 \end{pmatrix} \geq 0 \,. \tag{4.10.37}$$

Inequality (4.10.37) holds for all t in [0,1] so that the Riccati equation associated with the parameter values given in (4.10.28)-(4.10.30) has a solution for all t in [0,1], despite the fact that the conditions of Theorem 4.9 are violated.

4.11 Conclusion

In this chapter we studied rather thoroughly the quadratic functional that arises as the second variation in optimal control problems. First, we proved certain necessary and sufficient conditions for strong positivity of the nonsingular second variation and illustrated by means of an example that the totally singular second variation cannot be strongly positive. Next, we presented and proved a general sufficiency theorem for the partially singular case and specialized this in both nonsingular and totally singular cases. Following a detailed progression of lemmas we proved our main theorem in Section 4.6, providing necessary and sufficient conditions for non-negativity of the singular second variation. In the next section we deduced certain well known necessary conditions for optimality in singular control problems. Other necessary and sufficient conditions not derived in this chapter were then referred to and briefly discussed and the question of sufficient conditions for both a weak and a strong minimum were examined. Finally, we showed that the conditions of Theorem 4.2 allow us to deduce conditions for existence of a solution to the (nonsingular) matrix Riccati differential equation. These conditions are less stringent than those known heretofore.

References

Anderson, B. D. O. (1973). Partially Singular Linear-Quadratic Control Problems, IEEE Trans. autom. Control AC-18, 407-409.

Breakwell, J. V. and Ho, Y. C. (1965). On the Conjugate Point Condition for the Control Problem, Int. J. Engng. Sci. 2, 565-579.

Brockett, R. W. (1970). "Finite Dimensional Linear Systems". John Wiley & Sons, Inc., New York.

Coddington, E. A. and Levinson, N. (1955). "Theory of Ordinary Differential Equations". McGraw-Hill, New York.

Gabasov, R. (1968). Necessary Conditions for Optimality of Singular Control, Engng. Cybern. No.5, 28-37.

Gelfand, I. M. and Fomin, S. V. (1963). "Calculus of Variations". Prentice-Hall, Englewood Cliffs, N.J.

Goh, B. S. (1966). The Second Variation for the Singular Bolza Problem, SIAM J. Control 4, 309-325.

Goh, B. S. (1970). A Theory of the Second Variation in Optimal Control, unpublished report, Division of Applied Mechanics, Univ. California, Berkeley.

Jacobson, D. H. (1969). A New Necessary Condition of Optimality for Singular Control Problems, SIAM J. Control 7, 578-595.

Jacobson, D. H. (1970a). Sufficient Conditions for Nonnegativity of the Second Variation in Singular and Nonsingular Control Problems, SIAM J. Control 8, 403-423.

Jacobson, D. H. (1970b). New Conditions for Boundedness of the Solution of a Matrix Riccati Differential Equation, J. Differential Equations 8, 258-263.

Jacobson, D. H. (1971a). A General Sufficiency Theorem for the Second Variation, J. math. Analysis Applic. 34, 578-589.

Jacobson, D. H. (1971b). Totally Singular Quadratic Minimization Problems, IEEE Trans. autom. Control AC-16, 651-658.

Jacobson, D. H. and Speyer, J. L. (1971). Necessary and Sufficient Conditions for Optimality for Singular Control Problems: A Limit Approach, J. math. Analysis Applic. 34, 239-266.

Johansen, D. E. (1966). Convergence Properties of the Method of Gradients, in "Advances in Control Systems" (C. T. Leondes, ed.) Vol. 4, pp. 279-316. Academic Press, New York and London.

Kalman, R. E. (1960). Contributions to the Theory of Optimal Control, Bol. Soc. Mat. Mexicana 5, 102-119.

Kelley, H. J., Kopp, R. E. and Moyer, H. G. (1967). Singular Extremals in "Topics in Optimization" (G. Leitmann, ed.) pp. 63-101. Academic Press, New York.

McDanell, J. P. and Powers, W. F. (1970). New Jacobi-Type Necessary and Sufficient Conditions for Singular Optimization Problems, AIAA J. 8, 1416-1420.

Moyer, H. G. (1973). Sufficient Conditions for a Strong Minimum in Singular Control Problems, SIAM J. Control 11, 620-636.

Robbins, H. M. (1967). A Generalized Legendre-Clebsch Condition for the Singular Cases of Optimal Control, IBM Jl Res. Dev. 3, 361-372.

Speyer, J. L. and Jacobson, D. H. (1971). Necessary and Sufficient Conditions for Optimality for Singular Control Problems; A Transformation Approach, J. math. Analysis Applic. 33, 163-187.

Tait, K. S. (1965). Singular Problems in Optimal Control, Ph.D. dissertation, Harvard Univ., Cambridge, Mass.

CHAPTER 5

Computational Methods for Singular Control Problems

5.1 Introduction

It turns out that computational methods for singular optimal control problems have not been fully developed and there remains, therefore, considerable opportunity for further, but rather difficult, research in this area.

In this chapter, we survey certain numerical methods which have appeared during the past several years. First, we look at the problem of constructing a matrix function $P(\cdot)$ which satisfies the differential inequalities of Theorem 4.2 and of Theorem 4.5. This is an important numerical problem as these theorems state only that if a $P(\cdot)$ exists, then the second variation is non-negative; there is no hint as to how to construct $P(\cdot)$. Next, we look at the problem of computing controls which are, possibly, singular. In nonsingular situations there are numerous algorithms available, for example Jacobson and Mayne (1970), which iteratively improve a nominal, guessed, control function. However, in the singular case, most of these algorithms are not usable owing to the fact that certain steps in the algorithms become undefined. The remaining algorithms which are able to handle the singular case are of the first-order, or gradient, type, and these have been shown to behave particularly poorly, as far as rate of convergence is concerned, when applied to singular optimal control problems (Johansen, 1966). We therefore present a method (Jacobson et al., 1970) which converts a singular problem into a sequence of nonsingular ones by the addition of a term

$$\varepsilon \int_{t_o}^{t_f} u^T u dt.$$

This enables us to use powerful second-order algorithms (Jacobson and Mayne, 1970). As ε is progressively reduced toward zero we find that the solution of the ε-problem tends to that of the original, possibly singular, one. In this connection we refer to Powers and McDanell (1971) who have successfully used this technique in the singular Saturn guidance problem.

A recent generalized gradient method due to Mehra and Davis (1972) is then mentioned as are certain recent experiments using function space versions of Davidon's and Broyden's parameter optimization methods (Edge and Powers, 1974a, 1974b).

We then turn our attention to the singular linear-quadratic optimal control problem and give recent, relevant, references (Jacobson, 1972; Moore, 1969; Moylan and Moore, 1971; Wonham and Johnson, 1964). Finally, we emphasize certain work by McDanell and Powers (1971) and Maurer (1974) who have obtained interesting conditions which increase understanding of junctions between singular and nonsingular arcs.

5.2 Computational Application of the Sufficiency Conditions of Theorems 4.2 and 4.5

In this section we investigate certain methods for constructing a matrix function P(·) which satisfies the conditions of Theorem 4.2 and/or Theorem 4.5.

5.2.1 The Nonsingular Case

In the nonsingular case it is known, and proved in Theorem 4.3, that if S(·) exists which satisfies

$$-\dot{S}=Q+SA+A^TS-(C+B^TS)^TR^{-1}(C+B^TS) \qquad (5.2.1)$$

$$S(t_f) = Q_f \qquad (5.2.2)$$

then $P(\cdot) = S(\cdot)$ satisfies the conditions of Theorem 4.2. Furthermore, see Theorem 4.1, the existence of $S(t)$, $t_o < t \leq t_f$ is also a necessary condition for non-negativity of the second variation so that necessity of the conditions of Theorem 4.2 in the nonsingular case is implied. Thus it turns out that, in the nonsingular case, integration of (5.2.1), which can be accomplished numerically, is an altogether satisfactory way of constructing P(·).

5.2.2 The Totally Singular Case

If we are prepared to make the assumption that

$$(-1)\frac{\partial}{\partial u}\ddot{H}_u > 0, \qquad \text{for all } t \text{ in } [t_o, t_f] \qquad (5.2.3)$$

we have, from Theorem 4.4, that the existence of a function $S(\cdot)$ which satisfies, for all t in $[t_o,t_f]$, the equations

$$-\dot{S}=Q+SA+A^TS+[(AB-\dot{B})^TS+B^TQ-CA-\dot{C}]^T$$

$$[\frac{\partial}{\partial u}\ddot{H}_u]^{-1}[(AB-\dot{B})^TS+B^TQ-CA-\dot{C}] \tag{5.2.4}$$

$$C(t_f) + B^T(t_f)S(t_f) = 0 \tag{5.2.5}$$

$$Q_f-S(t_f) = 0 \tag{5.2.6}$$

implies that $P(\cdot)=S(\cdot)$ satisfies the conditions of Theorem 4.2. This is a Riccati differential equation but, unlike (5.2.1), equation (5.2.4) constructs an $S(\cdot)$ which satisfies

$$C+B^TS = 0 , \qquad \text{for all t in } [t_o,t_f]. \tag{5.2.7}$$

Again, see McDanell and Powers (1970), existence of $S(\cdot)$ turns out to be a necessary condition for non-negativity of the second variation so that the conditions of Theorem 4.2 are also necessary, subject to condition (5.2.3), for non-negativity of the second variation. The necessity of the existence of $S(\cdot)$ is obtained by means of Goh's transformation technique in (McDanell and Powers, 1970).

The fact that the S(·) which satisfies (5.2.4) - (5.2.6) also satisfies (5.2.7) raises the question of whether or not, in the singular case, there exists a Riccati differential equation of lower dimension, of the same form as (5.2.1). That this is indeed true is demonstrated by Speyer and Jacobson (1971) using Kelley's transformation technique. It is this type of reduction of the dimension of the state space to produce a nonsingular second variation which is exploited by Anderson and Moylan (1973, 1974).

If (5.2.3) is not satisfied then the singularity is deeper and one of the higher-order generalized Legendre-Clebsch conditions must turn out to be satisfied with strict inequality if a Riccati-approach to the construction of P(·) is to be used. It is therefore clear from these remarks that a method for constructing P(·) which is general, in the sense that it does not depend upon (5.2.3), is preferable. Such a method is discussed next.

5.2.3 A Limit Approach to the Construction of P(·)

The proof of Theorem 4.5 is based on the fact that the singular second variation can be made nonsingular by addition of the term

$$\int \frac{1}{2\varepsilon} u^T u dt$$

and that ε can be progressively increased so that

$$\lim_{\varepsilon\to\infty} S_\varepsilon(\cdot) = P(\cdot) \tag{5.2.8}$$

where

$$-\dot{S}_\varepsilon = Q+S_\varepsilon A+A^T S_\varepsilon-(C+B^T S_\varepsilon)^T(C+B^T S_\varepsilon)\varepsilon \tag{5.2.9}$$

$$S_\varepsilon(t_f) = Q_f \; . \tag{5.2.10}$$

Computationally, constructing $P(\cdot)$ according to (5.2.8) has both an advantage and a disadvantage. The disadvantage is rather obvious; namely, as $\varepsilon\to\infty$, the right hand side of (5.2.9) becomes ill defined. However, possible numerical difficulties can be reduced by solving (5.2.9) iteratively using a Newton or quasi-linearization method in which the solution for $\varepsilon=\varepsilon_1$ is used as a starting solution for the solution for $\varepsilon=\varepsilon_2>\varepsilon_1$. This is, in principle, a more stable approach than simply setting ε to a very large value and integrating (5.2.9) backwards from t_f using standard numerical integration techniques.

The advantage of using (5.2.8) is as follows: Often, we are not interested in the actual value of $P(\cdot)$ but rather whether or not a $P(\cdot)$ exists which satisfies the conditions of Theorem 4.5. In other words we may wish to construct $P(\cdot)$ only to verify its existence. Now, if no such $P(\cdot)$ exists we have from the lemmas preceding the proof of Theorem 4.5, and because of the continuity of $S_\varepsilon(\cdot)$ with respect to ε, that for some ε sufficiently large *but finite*, $S_\varepsilon(t)$ will cease to exist for some t in $(t_o, t_f]$. Thus the non-existence of a $P(\cdot)$ which satisfies the conditions of Theorem 4.5 can be inferred from the non-existence of $S_\varepsilon(\cdot)$, $\varepsilon<\infty$, ε sufficiently large. In other words, if no $P(\cdot)$ exists, it is not necessary to compute $\lim_{\varepsilon\to\infty} S_\varepsilon(\cdot)$ to confirm this; rather the integration of (5.2.9) may be halted as soon as a finite escape time is detected in $S_\varepsilon(\cdot)$.

To our knowledge, this limit approach for constructing, or testing the existence of, $P(\cdot)$ is the only one which is independent of assumptions such as (5.2.3).

5.3 Computation of Optimal Singular Controls

We now come to the question of synthesis of optimal control functions for, possibly, singular control problems. Of course, if it is known a-priori that the optimal control function has singular sub-arcs, and if the number and location of these sub-arcs are known, then special techniques can be devised to deal with this situation (Anderson, 1972). However, if little a-priori knowledge is at hand, an algorithm which can cope with singular sub-arcs, as and if these arise, is most attractive.

5.3.1 Preliminaries

We shall consider the problem of controlling

$$\dot{x}=f_1(x,t) + f_u(x,t)u; \quad x(t_o) = x_o \tag{5.3.1}$$

where x(t) is in R^n, u(t) is in R^m and where u(t) is constrained in the following way:

$$|u_i(t)| \leq 1 \quad \text{for all } t \text{ in } [t_o,t_f], \quad i=1,\ldots,m. \tag{5.3.2}$$

The performance index, or cost functional, is

$$J[u(\cdot)] = \int_{t_o}^{t_f} L(x,t)dt + F[x(t_f)] \tag{5.3.3}$$

where t_o and t_f are given. The functions $f_1(x,t)$, $f_u(x,t)$, L(x,t) and $F[x(t_f)]$ are assumed to be at least once continuously differentiable in each argument.

Our problem is to choose u(·), piecewise continuous in time, which satisfies (5.3.2), to minimize J[u(·)].

It is well known that, in the absence of certain further assumptions, the optimal control function for this class of problems consists of bang-bang and singular sub-arcs. Of course, if the optimal control is purely bang-bang (i.e. nonsingular) the techniques of Jacobson and Mayne, (1970) can handle the problem.

5.3.2 An ε-Algorithm

We develop here an algorithm for solving the

problem formulated in Section 5.3.1 which is applicable whether or not singular sub-arcs exist in the optimal control. Basically, we convert (5.3.3) into a non-singular functional by the addition of a term

$$\frac{\varepsilon_k}{2}\int_{t_o}^{t_f} u^T u dt.$$

Minimizing this nonsingular functional, using for example the techniques described by Jacobson and Mayne (1970), for a sequence of ε_k's such that $\lim_{k\to\infty} \varepsilon_k = 0$ yields the optimal value of (5.3.3). No assumptions as to the number and position of singular sub-arcs need be made.

First we define

$$J[u(\cdot),\varepsilon_k] = \int_{t_o}^{t_f}[L(x,t) + \frac{\varepsilon_k}{2}u^T u]dt + F[x(t_f)] \quad (5.3.4)$$

where

$$\varepsilon_k > 0 \quad (5.3.5)$$

Clearly this ε_k-problem is nonsingular and can be solved using the methods of Jacobson and Mayne (1970). Our ε-Algorithm is then as follows:

Step 1. Choose a starting value $\varepsilon_1 > 0$ and a nominal control function $\overline{u}_1(\cdot)$.

Step 2. Solve the resulting ε_k-problem (k=1 initially) using the methods of Jacobson and Mayne (1970); this yields a minimizing control function $u_k(\cdot)$.

Step 3. Choose $\varepsilon_{k+1} < \varepsilon_k$ (for example, $\varepsilon_{k+1}=\varepsilon_k/10$), set $\bar{u}_{k+1}(\cdot)=u_k(\cdot)$, $k=k+1$, and go to Step 2.

In a practical case the algorithm can be halted, after Step 2, when

$$\varepsilon_k < \sigma \tag{5.3.6}$$

where σ is a small, positive, pre-determined number.

We prove the convergence of the ε-Algorithm under the following assumptions.

Assumption 5.1 Let U be the set of piecewise continuous m-vector functions of time defined on $[t_o,t_f]$ which satisfy (5.3.2). Then

$$\inf_{u(\cdot)} J[u(\cdot)] = \min_{u(\cdot)} J[u(\cdot)] = \nu_o \tag{5.3.7}$$

where $u(\cdot)$ belongs to U and ν_o belongs to R^1.

Assumption 5.2

$$\inf_{u(\cdot)} J[u(\cdot),\varepsilon_k] = J[u_k(\cdot),\varepsilon_k] \text{ is obtained in U.}$$

That is, $u_k(\cdot)$, the control which minimizes $J[u(\cdot),\varepsilon_k]$ belongs to U.

These two assumptions allow us to prove the following lemma.

Lemma 5.1 For $k>\ell(\varepsilon_k<\varepsilon_\ell)$ we have

$$J[u_k(\cdot),\varepsilon_k] \le J[u_\ell(\cdot),\varepsilon_\ell]. \tag{5.3.8}$$

That is, $J[u_k(\cdot),\varepsilon_k]$ is a monotonically decreasing function of ε_k. Moreover, we have that

$$J[u_k(\cdot),\varepsilon_k] \ge \min_{u(\cdot)} J[u(\cdot)] = \nu_o, \tag{5.3.9}$$

$u(\cdot)$ belongs to U.

Proof: We have that

$$J[u_k(\cdot),\varepsilon_k] = \min_{u(\cdot)} J[u(\cdot),\varepsilon_k] \le J[u_\ell(\cdot),\varepsilon_k]$$

$$\le J[u_\ell(\cdot),\varepsilon_\ell] . \tag{5.3.10}$$

In addition,

$$\min_{u(\cdot)} J[u(\cdot),\varepsilon_k] = \min_{u(\cdot)} \{J[u(\cdot)] + \frac{\varepsilon_k}{2}\int_{t_o}^{t_f} u^T u dt\} \ge \nu_o. \tag{5.3.11}$$

We now state and prove our main theorem.

<u>Theorem 5.1</u> For a positive sequence $\{\varepsilon_k\}$, $\varepsilon_k>\varepsilon_{k+1}>0$ and $\lim_{k\to\infty} \varepsilon_k=0$, and under Assumptions 5.1, 5.2,

$$\lim_{k\to\infty} J[u_k(\cdot),\varepsilon_k] = \nu_o. \tag{5.3.12}$$

Proof: Since, by Lemma 5.1, $\{J[u_k(\cdot),\varepsilon_k]\}$ is monotone decreasing and bounded below by ν_o, it must converge.

Suppose, then, that

$$\lim_{k\to\infty} J[u_k(\cdot),\varepsilon_k] = \bar{\nu} > \nu_o. \tag{5.3.13}$$

But then, because of (5.3.2) and the fact that $-\infty<t_o<t_f<\infty$, we have for some ε_ℓ belonging to $\{\varepsilon_k\}$, ε_ℓ sufficiently small, that

$$J[u^o(\cdot),\varepsilon_\ell] < \bar{\nu} \tag{5.3.14}$$

where $u^o(\cdot)$ is defined by

$$\min_{u(\cdot)} J[u(\cdot)] = J[u^o(\cdot)] = \nu_o,\ u(\cdot) \text{ in } U. \tag{5.3.15}$$

Inequality (5.3.14) then implies that

$$J[u_\ell(\cdot),\varepsilon_\ell] < \bar{\nu}. \tag{5.3.16}$$

But this implies, since $\{J[u_k(\cdot),\varepsilon_k]\}$ is a monotone decreasing sequence that

$$\lim_{k\to\infty} J[u_k(\cdot),\varepsilon_k] < \bar{\nu}\ . \tag{5.3.17}$$

Hence, we have a contradiction and so (5.3.12) must be true.

It can happen that, as ε_k approaches zero, numerical ill-conditioning manifests itself. This is especially the case if second-order numerical methods are used, as then terms involving $\frac{1}{\varepsilon_k}$ occur (Jacobson et al., 1970). However, it must be acknowledged that "a sufficiently good approximation" to the solution can often be obtained by reducing ε_k to a small, but still numerically stable, value. Indeed, Powers and McDanell (1971) report a successful application of the ε-Algorithm to the important Saturn guidance problem.

5.3.3 A Generalized Gradient Method for Singular Arcs

Recently, Mehra and Davis (1972) exploited the observation that the independent variables in a computational procedure for determining optimal control do not have to be only the control variables. They show that a judicious mixture of control *and* state variables as independent variables can result in the control problem being nonsingular in these variables, and hence standard algorithms such as gradient or conjugate gradient can be used to advantage. Though various, relatively simple, problems are solved by Mehra and Davis (1972) using this idea, it seems to us that the procedure is not yet systematized to a point where it can be programmed into a package and used routinely. Thus it appears that additional research, in developing further these ideas, is necessary.

5.3.4 Function Space Quasi-Newton Methods

Perhaps the most encouraging numerical experiments in the computation of optimal (singular) controls are those of Edge and Powers (1974a, 1974b). The approach here is to use the powerful Davidon and Broyden parameter optimization methods by suitably extending these to function space. The rate of convergence experienced when using these methods is considerably faster than in the case of gradient and conjugate gradient methods, and the function space quasi-Newton methods are simpler to program than full second-order algorithms. In addition, only gradient information needs to be calculated, whereas second derivatives of the dynamic system and performance integrand need to be obtained explicitly in the case of second-order algorithms.

5.3.5 Outlook for Future

It is clear from the above sub-sections that much research needs to be done into the problem of computation of optimal (singular) controls. This is emphasized by the fact that the most promising experiments (Edge and Powers, 1974a, 1974b) are only of recent vintage.

5.4 Joining of Optimal Singular and Nonsingular Sub-Arcs

In this book our attention has been focussed mainly on problems totally singular for all t in $[t_o, t_f]$. However, practical problems usually exhibit a mixture, or concatenation, of nonsingular and singular sub-arcs in their optimal control functions. Here we give a brief survey of results in this area.

Study of junction conditions in linear-quadratic problems, where condition (5.2.3) is satisfied, has yielded a rather complete picture (Jacobson, 1972; Moore, 1969; Moylan and Moore, 1971; Wonham and Johnson, 1964), but in higher-order singular problems, junctions between singular and nonsingular segments are not entirely understood. It is fair to say, however, that McDanell and Powers (1971) have provided numerous insights into junction conditions via a number of theorems, one of which we state below without proof. It is important to note that the work of McDanell and Powers was stimulated by the pioneering contributions of Kelley et al., (1967).

Theorem 5.2 Let t_c be a point at which singular and nonsingular sub-arcs of an optimal control $u(\cdot)$ are joined, and let q be the order of the singular arc. Suppose that the strengthened, generalized Legendre-Clebsch necessary condition is satisfied at t_c; i.e.

$$(-1)^q \frac{\partial}{\partial u} [H_u^{(2q)}] > 0 \qquad (5.3.18)$$

and assume that the control is piecewise analytic in a neighbourhood of t_c. Let $u^{(r)}$ $(r \geq 0)$ be the lowest order time derivative of u that is discontinuous at t_c. Then, $q+r$ is an odd integer.

Corollary 5.1 In q-even problems, assuming u is piecewise analytic and the strengthened, generalized Legendre-

Clebsch condition is satisfied, the optimal control is continuous at each junction. See Maurer (1974) for an example.

Corollary 5.2 In q-odd problems, assuming u is piecewise analytic, and the strengthened, generalized Legendre-Clebsch condition is satisfied, the optimal control either has a jump discontinuity at each junction, or else the singular control joins the boundary smoothly; i.e. with a continuous first derivative.

5.5 Conclusion

In this Chapter we first examined the question of the construction of matrix functions of time which satisfy certain (necessary and) sufficient conditions for non-negativity of the second variation. It turns out, as is well known, that integration of the conventional matrix Riccati differential equation accomplishes this construction in the nonsingular case. In the totally singular case, under certain assumptions, a new Riccati differential equation is used in this role. In general, however, in the absence of these assumptions, the construction of the required matrix functions of time is not routine. The suggested method - a limit approach - while having certain numerical difficulties associated with it, appears to be the most useful general technique available.

Next, we considered the problem of computation of optimal singular controls. This is, in many respects, a similar problem to that discussed above in that the most flexible technique available turns out to be the ε-Algorithm, which is also a limit method. However, recent experiments using Davidon and Broyden algorithms in function space are very encouraging.

Finally, we discussed the joining of singular and nonsingular sub-arcs. In the linear-quadratic problem, under the assumption that the generalized Legendre-Clebsch condition is satisfied for q=1 with strict inequality, the situation is rather well understood. However, many difficult questions remain in the case of higher-order singular arcs.

We conclude, then, by noting that it is rather clear that further research needs to be done on computational methods in general and on computation of optimal singular controls in particular.

References

Anderson, B. D. O. and Moylan, P. J. (1973). Spectral Factorization of a Finite-Dimensional, Nonstationary Matrix Covariance, Tech. Rep. EE-7303, Univ. Newcastle, Australia.

Anderson, B. D. O. and Moylan, P. J. (1974). Synthesis of Linear Time-Varying Passive Networks, IEEE Trans. Circuits and Systems CAS-21, 678-687.

Anderson, G. M. (1972). An Indirect Numerical Method for the Solution of a Class of Optimal Control Problems with Singular Arcs, IEEE Trans. autom. Control AC-17, 363-365.

Edge, E. R. and Powers, W. F. (1974a). Shuttle Ascent Trajectory Optimization with Function Space Quasi-

Newton Techniques, AIAA Mechanics and Control of Flight Specialist Conference, Anaheim, California.

Edge, E. R. and Powers, W. F. (1974b). Function Space Quasi-Newton Algorithms for Optimal Control Problems with Bounded Controls and Singular Arcs, unpublished paper, Univ. Michigan, Ann Arbor.

Jacobson, D. H. (1972). On Singular Arcs and Surfaces in a Class of Quadratic Minimization Problems, J. math. Analysis Applic. 37, 185-201.

Jacobson, D. H. and Mayne, D. Q. (1970). "Differential Dynamic Programming", Elsevier, New York.

Jacobson, D. H., Gershwin, S. B. and Lele, M. M. (1970). Computation of Optimal Singular Controls, IEEE Trans. autom. Control AC-15, 67-73.

Johansen, D. E. (1966). Convergence Properties of the Method of Gradients, in "Advances in Control Systems" (C. T. Leondes, ed) Vol.4, pp.279-316. Academic Press, New York.

Kelley, H. J., Kopp, R. E. and Moyer, H. G. (1967). Singular Extremals, in "Topics in Optimization" (G. Leitmann, ed) pp.63-101. Academic Press, New York.

Maurer, H. (1974). An Example of a Continuous Junction for a Singular Control Problem of Even Order, Department of Mathematics, Univ. British Columbia, Vancouver, Canada.

McDanell, J. P. and Powers, W. F. (1970). New Jacobi-Type Necessary and Sufficient Conditions for Singular Optimization Problems, AIAA J. 8, 1416-1420.

McDanell, J. P. and Powers, W. F. (1971). Necessary Conditions for Joining Optimal Singular and Nonsingular Subarcs, SIAM J. Control 9, 161-173.

Mehra, R. K. and Davis, R. E. (1972). A Generalized Gradient Method for Optimal Control Problems with Inequality Constraints and Singular Arcs, IEEE Trans. autom. Control AC-17, 69-79.

Moore, J. B. (1969). A Note on a Singular Optimal Control Problem, Automatica 5, 857-858.

Moylan, P. J. and Moore, J. B. (1971). Generalizations of Singular Optimal Control Theory, Automatica 7, 591-598.

Powers, W. F. and McDanell, J. P. (1971). Switching Conditions and a Synthesis Technique for the Singular Saturn Guidance Problem, J. Spacecraft Rockets 8, 1027-1032.

Speyer, J. L. and Jacobson, D. H. (1971). Necessary and Sufficient Conditions for Optimality for Singular Control Problems; A Transformation Approach, J. math. Analysis Applic. 33, 163-187.

Wonham, W. M. and Johnson, C. D. (1964). Optimal Bang-Bang Control with Quadratic Performance Index, Trans. ASME, J. Basic Eng. 86, 107-115.

CHAPTER 6

Conclusion

6.1 The Importance of Singular Optimal Control Problems

The word "singular" in the context of optimal control theory has its origins in the classical calculus of variations in which a variational problem is referred to as singular if the Legendre-Clebsch necessary condition is not satisfied with strict inequality. In contemporary control theory we are consistent with this definition of singularity in referring to the second variation as totally singular if

$$H_{uu} = 0 \qquad \text{for all } t \text{ in } [t_o, t_f], \qquad (6.1.1)$$

as partially singular if

$$H_{uu} \geq 0 \qquad \text{for all } t \text{ in } [t_o, t_f], \qquad (6.1.2)$$

and as nonsingular if

$$H_{uu} > 0 \qquad \text{for all } t \text{ in } [t_o, t_f]. \qquad (6.1.3)$$

Thus, singular optimal control problems are "singular" in a mathematical sense; i.e. $\det(H_{uu}) = 0$. It turns out, though, that singular problems arise frequently in engineering and are, therefore, not at all "singular" or uncommon in the realm of applied optimal control theory. It was this realization, which probably began with Lawden's rocket problem, which stimulated the large research effort on these problems during the 1960's and early 1970's and which still motivates us to develop efficient computational techniques.

The importance of singular optimal control problems is therefore evident. On the one hand, singular problems exhibit many interesting and deep theoretical niceties and on the other hand they arise, and are therefore of practical significance, in engineering and other disciplines. Our aim in this monograph, therefore, is to give both sides of the story by referring in Chapters 1-3 to certain of the singular problems of practical importance, in Chapter 5 to computational techniques for their solution, and in Chapters 2-4 to necessary, and necessary and sufficient conditions for optimality. In the remainder of this chapter we discuss certain aspects of the results presented in Chapters 1-5 in order to stress their rather central theoretical and practical roles.

6.2 Necessary Conditions

It is well known that the applicability of Pontryagin's Principle is unaffected by singularity, or lack thereof. However, in the (partially) singular case H_{uu} is not invertible and therefore the sufficient conditions developed especially for the nonsingular case are inapplicable. This lack of sufficient conditions meant that the optimality, or lack thereof, of Lawden's spiral in the aerospace field could not be checked and this stimulated researchers in the quest for further necessary conditions. Research then yielded the generalized Legendre-Clebsch necessary condition of Goh, Kelley et al., and Robbins, which decided the fate of Lawden's spiral. It is rather interesting to note that Kelley et al. utilized special control variations of complicated nature to yield the generalized Legendre-Clebsch condition, but these researchers neglected to investigate further the affect of the simplest variation - a rectangular pulse. Presumably, they reasoned that this simple variation, which yields the classical Legendre-Clebsch condition in the nonsingular and partially singular cases, could not yield any additional useful information. Remarkably, use of this simple variation does yield additional results as evidenced by Jacobson's necessary condition which emerged circa 1969 and which paved the way for the development of necessary and sufficient conditions.

A further consequence of the search for additional necessary conditions was the development of Kelley's and Goh's transformation techniques. These, in turn, played an important role in the development of necessary and sufficient conditions.

6.3 Necessary and Sufficient Conditions

In the (partially) singular control problem the conventional Riccati differential equation is not defined, owing to the presence of H_{uu}^{-1} in its right hand side. For a long time, then, it was felt that no sufficiency theory analogous to that for nonsingular problems could be expected in the singular case. This is, in retrospect, a somewhat surprising conclusion to reach as Riccati-like inequalities were already known in network theory in connection with passivity which is closely related to the non-negativity of a singular quadratic functional.

Shortly after the advent of Jacobson's necessary condition a set of algebraic and differential inequalities, closely related to conditions in network theory, were developed which are sufficient for non-negativity of the singular second variation. At this point in time the quest for conditions which are both necessary and sufficient intensified. Two different approaches were used; namely, a limit approach of Jacobson and Speyer and the transformation approaches of Goh, Powers, and Speyer and Jacobson.

The limit approach of Jacobson and Speyer, fully detailed in Chapter 4, converts the singular second variation into a nonsingular one by the addition of the term

$$\frac{1}{2\varepsilon}\int_{t_o}^{t_f} u^T u dt.$$

The conventional Riccati equation for this nonsingular problem is then studied in the limit as $\varepsilon \to \infty$, yielding necessary and sufficient conditions for non-negativity of the original singular variation. This limit approach has proved to be rather powerful in that minimal differentiability assumptions are required. Furthermore, all known necessary conditions can be recovered by manipulating, in a rather simple way, these conditions. In certain cases it is possible to exhibit the necessary and sufficient conditions in Riccati differential equation form thereby demonstrating the close relationship between singular and nonsingular problems.

The transformation approaches are designed to transform the singular problem into a nonsingular one; the conventional necessary and sufficient conditions (conventional Riccati equation) are then applicable. In the Speyer and Jacobson approach, Kelley's transformation technique converts the singular problem into a nonsingular one in a state space of reduced dimension. Goh's transformation, on the other hand, as exploited by McDanell and Powers, preserves the dimensionality of the original state space. As Goh's transformation is simpler to apply than Kelley's we use it here to illustrate the role of transformation techniques in the development of the Riccati equation for singular problems which is stated and used, but not derived, in Chapter 4.

We consider the totally singular quadratic problem,

$$J[u(\cdot)] = \int_{t_o}^{t_f} (\tfrac{1}{2}x^T Qx + u^T Cx)\,dt + \tfrac{1}{2}x^T(t_f)Q_f x(t_f) \quad (6.3.1)$$

where

$$\dot{x} = Ax + Bu \; ; \quad x(t_o) = x_o . \quad (6.3.2)$$

Goh's transformation technique is as follows. First, define

$$v = \int_{t_o}^{t} u(\tau)\,d\tau \quad (6.3.3)$$

and

$$z = x - Bv \, . \quad (6.3.4)$$

Then, we obtain easily the differential equation,

$$\dot{z} = Az + (AB - \dot{B})v \; ; \quad z(t_o) = 0 \quad (6.3.5)$$

and

$$J[v(\cdot)] = \int_{t_o}^{t_f} (\tfrac{1}{2}z^T Qz + v^T B^T Qz + \tfrac{1}{2}v^T B^T QBv + \dot{v}^T Cz + \dot{v}^T CBv)\,dt$$

$$+ \tfrac{1}{2}[z(t_f) + Bv(t_f)]^T Q_f [z(t_f) + Bv(t_f)]$$

$$(6.3.6)$$

Integrating the terms in $\dot{v}$ by parts and re-arranging yields

$$J[v(\cdot)] = \int_{t_o}^{t_f} \{\tfrac{1}{2}z^T Qz + v^T(B^T Q - CA - \dot{C})z + \tfrac{1}{2}v^T[-\frac{\partial}{\partial u}\ddot{H}_u]v$$

$$+ \tfrac{1}{2}\dot{v}^T[CB - (CB)^T]v\}dt$$

$$+ \tfrac{1}{2}z^T(t_f)Q_f z(t_f) + v^T(t_f)[C + B^T Q_f]z(t_f)$$

$$+ \tfrac{1}{2}v^T(t_f)[CB + B^T Q_f B]v(t_f) \quad (6.3.7)$$

Now, if the first generalized Legendre-Clebsch necessary condition is satisfied in strong form, i.e.

$$CB = (CB)^T \quad (6.3.8)$$

and

$$-\frac{\partial}{\partial u}\ddot{H}_u > 0 \quad (6.3.9)$$

and if we assume that

$$C + B^T Q_f = 0 \qquad \text{at } t_f, \quad (6.3.10)$$

then,

$$J[v(\cdot)] = \int_{t_o}^{t_f} \{\tfrac{1}{2}z^T Qz + v^T(B^T Q - CA - \dot{C})z + \tfrac{1}{2}v^T[-\frac{\partial}{\partial u}\ddot{H}_u]v\}dt$$

$$+ \tfrac{1}{2}z^T(t_f)Q_f z(t_f) \quad (6.3.11)$$

which is a standard, nonsingular quadratic functional. A necessary and sufficient condition for strong positivity for (6.3.11) is, then, that there exists $S(\cdot)$ which satisfies

$$C+B^T S(t_f) = 0 \tag{6.3.12}$$

$$-\dot{S}=Q+SA+A^T S+(B^T Q-CA-\dot{C}+B^T A^T S-\dot{B}^T S)^T$$

$$\frac{\partial}{\partial u}\ddot{H}_u^{-1}(B^T Q-CA-\dot{C}+B^T A^T S-\dot{B}^T S) \tag{6.3.13}$$

where

$$S(t_f) = Q_f \,. \tag{6.3.14}$$

On the application of Kelley's transformation rather than Goh's, Speyer and Jacobson's Riccati equation which is of smaller dimension, but otherwise equivalent, is obtained.

If it turns out that the first generalized Legendre-Clebsch condition is not satisfied in strong form the transformation techniques have to be applied repeatedly until one of the higher-order generalized Legendre-Clebsch conditions is satisfied in strong form. This can only be done if the system equations and performance integrand are many times differentiable; a significant disadvantage which is not present in the more direct limit approach of Chapter 4.

6.4 Computational Methods

Computational problems in singular control fall rather naturally into two areas. The first is the computation of optimal singular controls and the second is the construction of certain matrix functions of time in necessary and sufficient conditions which check a particular singular control function for optimality. It turns out that the limit approach, in which the singular problem is made nonsingular by the additions of

$$\frac{1}{2\varepsilon}\int_{t_o}^{t_f} u^T u dt ,$$

is useful in both of these problem areas. Indeed, it is worth mentioning that Powers and McDanell have reported using this method successfully in a singular Saturn guidance problem. Details of the limit approach are given in Chapter 5 where it is pointed out that this is the only general method currently available. However, the recent experiments of Edge and Powers involving quasi-Newton methods in function space are encouraging and augur well for the future.

6.5 Switching Conditions

The joining of singular and nonsingular sub-arcs is important, and not yet fully understood. However, McDanell and Powers have provided a number of useful, general, theorems and Jacobson has obtained satisfactory switching strategies in a class of quadratic problems.

6.6 Outlook for the Future

It seems to us that the areas of optimal (singular) control which require greatest attention are those which involve the computation of optimal controls and the joining of singular and nonsingular segments. The experiments of Edge and Powers in control function computation and the work of Anderson and Moylan in network synthesis are encouraging.

In closing, we mention the recent work of Moyer in connection with sufficient conditions for a strong minimum in singular optimization problems. This paper could stimulate further developments in optimal control theory.

AUTHOR INDEX

SUBJECT INDEX